AN INTRODUCTION TO THE MATH OF VOTING METHODS

AN INTRODUCTION TO THE MATH OF VOTING METHODS

by

Brendan W. Sullivan

619 Wreath

First edition, 2022

ISBN 978-1-958469-03-3

Published by 619 Wreath

Acknowledgments

This book project grew from teaching about voting methods and elections in the course "Mathematical Reasoning for Modern Society" at Emmanuel College. About 10 years ago I knew nothing about these topics, but teaching that course inspired me to learn more about the mechanics of democracy and how the systems we have in place influence real world behavior and outcomes. Working with students has shown me that people are eager to learn about these important issues and have informed, productive discussions. I hope that this book can make a positive contribution to the growing national conversation about electoral mechanics, and any success towards that goal is thanks in large part to the many students who have taken that course with me. Beyond reading early versions of this text and making concrete suggestions, students have also helped me realize which examples best illustrate an idea, which challenging concepts require careful explanations, which real world topics can motivate learning about abstract ideas, and so much more. I offer my heartfelt thanks to all of these students: you have been far more helpful than you may even realize!

I would also like to acknowledge the support of my Emmanuel colleagues along this journey. It has been a wonderful opportunity to teach and redesign that course I mentioned above, and I am grateful that the Mathematics Department gave me that chance. I have also given presentations about these topics through the library and even wrote a blog post about ranked choice voting for the college website. Those opportunities, as well as conversations with faculty colleagues, gave me some much needed confidence and inspired me to pursue this project.

And, of course, I could not have completed this book without the support of family and friends. Many thanks to everyone in my life who has spoken with me about voting in the last few years, especially my love, Kim, who has endured more of those conversations than anyone else by far. Our daughter, Ruby Joan, has been too young to join those discussions, but I look forward to helping her learn about the importance of voting soon enough.

Finally, the folks at 619 Wreath Publishing have been wonderfully helpful through the editing process and I will be forever grateful for their support of this project. One thing I especially admire is their commitment to donate to good causes. Because of this, a portion of your purchase will support the following organizations:

- *Fair Fight:* It doesn't make sense to talk about voting methods without securing everyone's right to vote in the first place. This organization does important work to "ensure every American has an equal opportunity to make their voice heard and be fairly represented." [1]

- *The Young People's Project:* Voting methods is a topic that may not appear "mathematical" at first but certainly requires logical, problem-solving, and quantitative reasoning skills. This organization trains students to be Math Literacy Workers who "become engaged citizens prepared to make a difference in their own lives, in the lives of others in their communities, and ultimately in this country." [2]

And lastly, thank *you* for your interest in this book. Enjoy!

Preface

You may have some experience with voting in an election. Maybe you haven't participated yourself, but you are aware of elections happening and how they work. Or, maybe you've voted often and you encourage your friends and family members to vote, too. No matter what your experience, the goal of this book is to show you that **there are several different ways for voters to submit ballots and have those ballots counted.**

We will learn about several different **voting methods**. A voting method is a set of rules that tells us how to combine all of the information written on the voters' ballots and calculate a winner of an election. (The rules may also say how to determine a second place candidate, and so on, if needed.)

You likely already know a simple voting method: **Majority Voting**, or just "majority rules." If there are two options, every voter picks one of them, and you just count whichever one gets the most votes. That option wins because they must have more than half of all the votes; that's what a majority is, something more than 50%. (Okay, there could be a perfect tie. But that's pretty rare. And what to do when that happens is not a *mathematical* question; it depends on the context of the situation.)

If you vote in the United States, you are familiar with another voting method: **Plurality Voting**. You might not know the name, but you know its rules: no matter *how many* options there are, every voter picks one of them, and you just count whichever one gets the most votes. They might not have *more than half* of all the votes, but they have *more votes than the other options*. That's what a **plurality** is, a winning total that is less than 50% of all votes.

Almost every election in the United States uses the **Plurality Voting** method. In fact, it's so prevalent that we worry many people are not even aware that there are other ways to vote! Indeed, we're assuming you're one of those people, so if you happen to already know a little bit about *"alternative voting methods"* (like **Ranked Choice Voting**, which has been in the news), then you might recognize some concepts as you read. But, you are absolutely not at a disadvantage if this is all new to you; in fact, we're assuming that this is your first time thinking carefully and critically about *how* we vote.

And that leads to a concluding remark about our goals in this book: **We can discuss *how we vote* without necessarily getting into a debate about *who we vote for*.** That is, this book is about mathematics, not politics. We will use our critical thinking skills to better understand certain ideas about voting and elections. We will use logic to explain our observations and support our opinions. We will analyze data to better understand someone's ideas or support our own. And, to make these concepts tangible and underscore their importance, we will discuss actual examples from history and current events. In short, we cannot claim that we will "never talk about politics." But, please keep in mind that politics is not the *focus* of this book. We would like to help you develop important knowledge and reasoning skills so that when you go out there in the world and discuss voting and elections, you will be able to contribute to informed and productive debates.

Contents

Chapter 1:
Majority Voting and Plurality Voting ("First Past the Post")

THE goal of this chapter is to introduce you to **Plurality Voting** (also known as **"First Past the Post"**), which is common and popular around the world. We will describe how it works and where it is used, and briefly discuss some of its pros and cons. Along the way, we will introduce some important concepts and examples that will be mentioned throughout the rest of the book.

1.1 Definitions and Examples

Let's begin with the simplest **voting method**, or set of rules for calculating a winner of an election.

Definition 1.1: Majority Rules Voting.
*This is a voting method that only applies when there are two candidates. Each voter chooses one of the two candidates. The winner is the one that receives the most votes, which must be a **majority** (more than half, $> 50\%$).*

If there is an even number of voters, then it is possible for the result to be a perfect tie, so the organizers of a real-world election may wish to have a "tie-breaker rule" in place. However, this is not a *mathematical* issue: it depends on the *context*. How to handle a tie in an election for a Senate seat might be very different from how you handle a tie in a friendly vote about what to have for dinner! (And with thousands or even millions of voters in a Senate race, a perfect tie would be astronomically unlikely, anyway.) So, when we look at examples in this book, we will not deal with ties all that often, and we will never presume any particular method for breaking a tie.

Example 1.1. Suppose we have nine friends who want to order a meal for a group study session. In a friendly discussion, they narrow down their options to Pizza or Indian food. They agree that everyone will write down either *"P"* or *"I"* on a slip of paper to indicate their choice, and then the **"Majority Rules" method** will be used to decide the winner.

Let's say that five friends chose P and the other four chose I. In that case, Pizza wins the election with 5 votes out of 9 total votes, or a proportion of $\frac{5}{9}$, which is approximately 55.6%.

This simple example brings up a few important ideas:

- In this example, the **candidates** are the options to choose from: Pizza and Indian food. In some examples, the candidates will be people running for political office, but this example shows that this is not always the case.

- In this example, the **voters** are the nine friends. In some examples, the voters may not even be people, but usually they will be.

- In any example you come across, in this book or elsewhere, it is probably worth spending a moment to think about who/what the candidates are, and who/what the voters are.

- In this example, a **ballot** submitted by one of the voters is just an indication of which option is that voter's top choice. Imagine that you're one of the friends and you want Indian food. By writing *"I"* on your slip of paper, you are indicating that as your top choice.

  If you had instead written "1st: *I*, 2nd: *P*," that would provide *the exact same information*, but with more writing. In other words, with only two candidates, you can just indicate which one is your 1st place choice; it is automatically implied that the other option is your 2nd place choice. This may seem silly to point out now, but very soon we will look at situations with three or more candidates, in which case a voter's ballot may contain much more information than merely saying, "Here is my #1 choice."

- Beyond what the ballots say, it is also worth considering *how the ballots are submitted*. What if the friends raised their hands to vote publicly? Perhaps some of them would "jump on the bandwagon" and vote for the popular choice because they see almost everyone else raising their hand already. Or, in a larger group, perhaps they would use some kind of online survey to gather the votes instead of using slips of paper.

  These are important considerations if you're going to hold an election or participate in one. However, you can see that they are not really *mathematical* issues; rather, they depend on the "real world" context of the election. For this reason, we will not really discuss such issues in this book. For the most part, we will focus on *how to take the information on the ballots and calculate the winning candidate*. All other considerations (how the candidates are chosen, what voters get to participate, how the ballots are submitted, etc.) are outside of our scope.

Let's now look at our first example of a voting method that applies when there are *more than two candidates*. Imagine the friends decide to expand their meal options: Pizza, Indian food, and Thai food. In this situation, "majority rules" may not apply because *the votes can be split among the three options, causing none of them to have $> 50\%$ support*. This is what we will see in the example below, after the definition.

Definition 1.2: Plurality Voting.
*This is a voting method that applies when there are three or more candidates. Each voter chooses one of the candidates. The winner is the one that receives the most votes. It is possible for the winner to have less than half of all the votes, in which case their share is called a **plurality**. (Remember that the term **majority** is only used when the winner has $> 50\%$ of all the votes.)*

Example 1.2. Suppose the nine friends decide that each of them will write *"I"* (Indian food) or *"P"* (Pizza) or *"T"* (Thai food) on a slip of paper. The votes will be counted, and the option with the most votes will be the winner. Let's say four friends choose I, three choose P, and two choose T. In other words,

- Indian food gets $\frac{4}{9}$ of the votes, or approximately 44.4%.

- Pizza gets $\frac{3}{9}$ of the votes, or approximately 33.3%.

- Thai food gets $\frac{2}{9}$ of the votes, or approximately 22.2%.

You can see that Indian food is declared the winner with a mere **plurality** (44.4%) and not a **majority** ($> 50\%$).

A few comments about this example:

- In the previous example with only two options, Pizza got five votes. In this new example here, it may help to imagine that two of the friends switched their choice from Pizza to Thai food when that option became available, while all four of the voters who picked Indian food originally stayed with that choice.

- This is just one example of a winner having a plurality but not a majority. We will see many examples of that in this book, and we will mention plenty of similar real world examples.

- We specifically chose the same underlying situation to use in these first two examples to make it clear that *Majority Rules is Plurality Voting when there are only two candidates.* In other words, these two methods are fundamentally identical. The only differences arise when the number of candidates is more than two.

1.2   Real World Usage

Plurality Voting is used all over the place; it's fair to say it's the most popular voting method, at least by frequency of use. If you have participated in pretty much any election in the United States, then you have used Plurality Voting.

You may have seen the phrase **"First Past the Post"** used, perhaps in a news report about voting. This phrase refers specifically to an election where Plurality Voting is used and there is a single winner. For example, every two years in the US, you get to vote for your representative in the U.S. House of Representatives. There may be several candidates, but you only get to indicate your support for *one of the candidates* on the ballot, and the single winner is the one who gets the most support.

Many nations around the world use a First Past the Post (Plurality) method when citizens vote for their representatives in the national legislature. This includes the U.S. (Senate and House of Representatives), the United Kingdom (Parliament), Brazil (Federal Senate), Canada (House of Commons), Azerbaijan (National Assembly), Ghana (Parliament), India (Lok Sabha, or House of the People), and many more. See the Wikipedia page on "First-past-the-post voting" for a list of countries that currently use such a method to elect their Head of State or National Legislature. [3]

You have likely also encountered Majority Rules Voting and Plurality Voting in school. Perhaps a teacher needed to decide what to do next, and they held a vote among the class. If there were only two options, they probably used the Majority Rules method. And if there were three or more options, they probably used the Plurality Voting method. We're guessing that this applies to your Student Government elections, too: voters get to *pick one* of the options (however many there are), and the winner is the one who gets the most votes.

In short, Plurality Voting is used all over the place because it is straightforward and simple to implement. However many candidates there are, each voter just picks one and indicates that choice somehow. Then, those choices are counted and you see which total is largest. Couldn't be simpler, right? This is one of the pros of the method that we will discuss shortly.

1.2.1   Real World Examples of Elections

Let's look at a couple of specific examples of political elections in the U.S. that used Plurality Voting. They have been chosen to point out some important ideas and contribute to some later discussions.

6

Example 1.3 (Massachusetts' 3rd Congressional District, 2018). The general election for all U.S. House of Representatives seats was held on November 6, 2018. In Massachusetts' 3rd District, there were three main candidates; here are their vote totals [4]:

- Lori Trahan (Democrat) won 173,175 votes, about 62% of the total.

- Rick Green (Republican) won 93,445 votes, about 33.5% of the total.

- Mike Mullen (Independent) won 12,572 votes, about 4.5% of the total.

- There were also 135 write-in votes for other candidates.

You can see that Lori Trahan won the election and she even had a **majority** of the votes. However, there was a much closer race two months earlier: the Democratic primary was held on September 4, but the results were so tight that a manual recount didn't conclude until two weeks later.

In that primary, Lori Trahan won with not a majority but rather a very slim **plurality**. There were ten candidates all vying for the Democratic spot on the ticket in the upcoming general election. The votes were pretty well spread across all ten candidates so that no one got even close to a majority. In fact, five candidates all earned somewhere between 15% and 22% of the votes, accounting for almost 90% of all votes (with the other five candidates accounting for the remaining 10% or so). Here are the results for the top six [4]:

- Lori Trahan won 18,580 votes, about 21.7% of the total.

- Dan Koh won 18,435 votes, about 21.5% of the total.

7

- Barbara L'Italien won 13,018 votes, about 15.21% of the total.

- Juana Matias won 12,993 votes, about 15.18% of the total.

- Rufus Gifford won 12,873 votes, about 15.04% of the total.

- Alexandra Chandler won 4,846 votes, about 5.7% of the total.

Wow! It's remarkable how close 1st and 2nd place were, as well as how close 3rd, 4th, and 5th all were. But more importantly, it's worth considering the fact that while 21.7% of the voters in that district chose Lori Trahan, 72.3% of the voters chose *someone else*. With 10 candidates, the votes can be spread out so much that the winner receives a rather small percentage of the total.

Now, we don't mean to cast doubt on Lori Trahan or the legitimacy of the election, not at all. Presumably, many of the the voters who chose someone other than her were totally fine with her winning. But what if a lot of them really hated her and would have ranked her last, if they could have put all the candidates in their order of preference? Wouldn't that be strange, to have such a widely unpopular candidate eke out a victory, and partly just because there were so many people running? Again, we are **not** saying this is actually what happened in that election, but it is important to recognize that such a thing could easily happen, *because of the rules of the Plurality Voting method.*

1.2.2 Moving Away from First Past the Post

Because a major goal of this book is to give you information with which to have an informed discussion about different voting methods, it's also worth mentioning now that several countries have recently *changed* from First Past the Post voting to something else.

For example, in 2003 Papua New Guinea changed from First Past the Post to something similar to **Instant Runoff Voting** (which we will learn about in the next chapter, and is also known as **Ranked Choice Voting**). Even more recently, Lebanon moved to a **proportional representation** system in 2018. Each **district** is represented by multiple members of the legislature, and voters in a district can vote for several candidates. The results of that district are based on the proportion of all votes that each candidate receives.

In the United States, Plurality Voting is still used in all elections for Congress and the President (although the Electoral College complicates the process somewhat), *except in Maine and Alaska.* In 2016, voters in Maine narrowly approved a ballot referendum that stipulates Ranked Choice Voting will be used in all Congressional elections from now on. The midterm election in 2018 was the first time a method other than Plurality Voting was used in a statewide election in the US. And voters in Alaska passed a ballot referendum in 2020 that will see them using a two-round combination of Plurality Voting and Ranked Choice Voting in the future. (All of this will be discussed later in Section 2.2.2.)

For now, we hope that you see that Plurality Voting is common and widespread, but also that it does not *have* to be so common. Let us now discuss some reasons for and against using this voting method.

1.3   Discussion of Pros and Cons

1.3.1   Pros

We've already mentioned the most obvious pro of Plurality Voting: *simplicity*. It's worth describing how this applies in a few different ways:

1. **Simplicity for the voters.** The instructions on a ballot are easy to state: *select one.* And those instructions are easy to follow: a voter fills in one bubble, or checks one box, or something like that. In other words, this method has a very low potential for voter error or confusion.

2. **Simplicity in calculating results.** Because each submitted ballot essentially has one piece of information (the one selected candidate), it's straightforward to count how many votes each candidate receives. This can become logistically challenging when there are millions of voters, but this method is, at least, mathematically very simple.

3. **Simplicity for the candidates.** Assume we're talking about a political race. When each candidate knows that *"the most votes wins"* and that's it, they will act accordingly in their campaigns. They will tell you (the voters) why to choose them and not the others. Although this is not a great system and tends to encourage negative campaigning, this is simple.

1.3.2 Cons

Although it may seem strange, one of the most popular arguments about the downside of Plurality Voting is that it is *too simple.* That is, all of those Pros above can be interpreted as Cons.

1. **Too simple for the voters: cannot express nuanced opinions.** With three or more options, voters may not want to just choose one, or they may wish that they could express their *order of preference.*

   For example, imagine you're one of the nine friends choosing to order Pizza, Indian, or Thai food. Maybe you're a big fan of Pizza and Thai food, and you really can't decide between them, but you really don't want Indian at all. How should you vote? There is no way to express that opinion using the Plurality Voting method! There is no way to indicate: "These two are tied for first, and this one is a distant third."

In the examples above, we hypothesized five voters choosing Pizza and four voters choosing Indian, when the choices were just those two. But when the option of Thai food was introduced, two of the Pizza supporters switched to Thai, and that ended up making Pizza *lose* the election! Wouldn't that be strange if all five of the original Pizza supporters (a majority of the electorate, mind you) absolutely hated Indian food, and yet that option *won the election* because the Pizza supporters split their vote amongst Pizza and Thai?

2. **Too simple as a system: discourages 3rd parties.** The scenario we just described — with Thai *"splitting the vote"* and causing Indian to win — is not just a hypothetical one. Studies of First Past the Post Voting systems in the real world have shown that the mechanics of the voting system *encourage and entrench a Two Party System.* That is, it is not just because of the voters, or the candidates, or the political activity of the country: it's the *method of voting* that seems to foster a system where two major parties dominate the political landscape. [5]

In the example above, Pizza and Indian are the major party options, with Pizza enjoying a slim majority position. But when 3rd party Thai tried to enter the mix, some supporters of Pizza jumped ship and gave rival Indian the victory! This kind of phenomenon is why you hear about the **spoiler effect** and **"throwing your vote away,"** especially when there are candidates from outside the Democrat/Republican major parties.

3. **Too simple for candidates: encourages negative campaigning.** This is, admittedly, a non-mathematical issue, but we will mention it for a couple of reasons. For one, this rounds out the Pro/Con lists of this section. For another, this will contrast directly with a significant Pro of the **Instant Runoff (Ranked Choice)** voting method that we will learn about in the next chapter.

When candidates are trying to encourage voters to choose them and no one else, this *tends to incentivize "attack ads."* That is, a candidate knows that a voter can only pick one option and they want to be chosen, so they might use their platform to explain to voters why *their opponents are terrible* instead of why *they are a good candidate*.

1.4 Practice Problems with Solutions

Each chapter of this book includes a section of example problems to help us reinforce essential knowledge and skills. We'll often phrase these problems in a way that you might see on a homework assignment, and we'll focus on explaining *how to solve the problems*, not just on what the answers are. For now, we have not introduced too much new information, so there's not a lot to do yet. So, let's just look at another example of an election and answer some questions about it to practice our critical-thinking and information literacy skills.

Practice Problem 1.1. This problem is based on real data from an election: the Democratic Primary held on September 4, 2018 for the Massachusetts State House of Representatives in the 11th Essex legislative district.

1. The following website has a list of the results from all primary elections held in Massachusetts on September 4, 2018. Find the data table showing the results of the specific election mentioned in this problem. [6]

   `https://bit.ly/MA180904`

2. Confirm the percentages listed in that data table.

3. Explain why those percentages don't add up to what you'd expect.

4. Was this a majority victory? Why or why not?

5. What if all the voters who chose the candidate in last place were given a chance to change their vote to one of the other two instead? Could that affect the outcome of the election? By how much?

6. What if all the voters who chose the candidate in last place stayed home instead and didn't vote? Could that affect the outcome of the election?

▼ Solution:

1. The problem statement mentioned the *"11th Essex"* district. Hit ctrl-F (*Find*) on that page to search for that text, and it will jump right to what we need. Here's a screenshot [6]:

State House - 11th Essex - Dem - Primary
12 of 12 Precincts Reporting - 100%   Updated: Sep. 04, 2018 9:58 pm EST

Party	Name	Votes	Vote %	
Dem	Capano, Peter ●	1,840	47.85%	
Dem	Russo, Drew	1,422	36.98%	
Dem	Net, Hong	583	15.16%	

2. *"Confirm the percentages"* means double-check the calculations. Before we do that, it's worth pointing out how you can usually do a simple *"spot check"* just to see if the numbers even make sense.

    In this example, we might estimate the total number of voters by adding $2,000 + 1,500 + 500 = 4,000$. That's not perfect, but it's pretty close and uses nice, round values to make the mental arithmetic easier. We can now see that Peter Capano's vote share is roughly 2,000 out of 4,000, which would be 50%, and so 47.85% sounds reasonable. Likewise, Hong Net's share is roughly 500 out of 4,000, or 1/8 (or 12%-ish), so 15% seems reasonable. And Drew Russo got roughly 500 votes less than Capano, so 1/8 (or 12%-ish) less than him, which is pretty close ($48 - 12 = 36$).

    That estimation is the kind of thing you can sometimes do in your head, but of course a calculator can always help, so let's do that, too:

    - Total votes: $1,840 + 1,422 + 583 = 3,845$
    - Capano's vote share: $\frac{1,840}{3,845} = 0.47854\ldots \approx 47.85\%$
    - Russo's vote share: $\frac{1,422}{3,845} = 0.36983\ldots \approx 36.98\%$
    - Net's vote share: $\frac{583}{3,845} = 0.15162\ldots \approx 15.16\%$

3. That accounts for all the votes, so those percentages *should* add to 100%. But ... $47.85 + 36.98 + 15.16 = 99.99$ Huh?

This is a matter of *rounding*. Capano's vote share is not exactly 47.85%. That may *feel like a very precise number* because it has two decimal places but it is not perfectly accurate. When we rounded 0.478543... down to 0.4785 (before converting that to 47.85%), *we made the number smaller.* Likewise, we rounded 0.369830... *down* to 0.3698 (Capano's share), and we rounded 0.151625... *down* to 0.1516 (Net's share). So it should not be surprising that the total is a little bit less than 100%.

Now, we would be concerned if the total were, say, 94.82%. That number is also less than 100%, but it's off by so much that it indicates a genuine arithmetic error, maybe even a typo when we pushed buttons on the calculator. You should always double-check calculations to be careful. (In this case, that meant adding the percentages to see if they total to 100%.) However, you should also be aware that small discrepancies may be accounted for by small roundoff values (essentially, "a little missing information"), but large discrepancies may indicate mistakes to be fixed.

4. The largest vote share was 47.85%, which is *not* more than half. So, this was merely a **plurality** victory, not a majority.

5. We see that Hong Net got 15.16% of the vote. This question is asking us to explore what would happen if that share of the vote could be redistributed among the other candidates.

   What if all of them voted for Peter Capano instead? Obviously, he would win by an even wider margin than he did before, but by how much? Would it be a majority?

   - Capano's original votes + Net's original votes:
     $1,840 + 583 = 2,423$

- Capano's new vote share:
  $\frac{2,423}{3,845} = 0.63016\ldots \approx 63.02\%$

- Russo's new vote share:
  $\frac{1,422}{3,845} = 0.36983\ldots \approx 36.98\%$

So, yes, Capano would win with a majority.

You might suggest that we could just add Capano's vote share + Net's vote share to get $47.85\% + 15.16\% = 63.01\%$. But notice there's a small roundoff error there, too! Keep this in mind: whenever you're working with numbers that have *already been rounded off*, if you do further calculations with those rounded values, there may be small discrepancies.

A more general lesson, as well: if you're working with *percentages*, the most accurate option is to always work with *the underlying information* before it has been turned into a proportion/percentage.

What if all of Hong Net's voters switched to Drew Russo instead? We bet that you can do the calculations to confirm that Russo's new vote share would be 52.15%. This means he would have a slim majority victory.

The point of asking these questions is to show you that the results could vary if the supporters of the "last place candidate" were allowed to retain their voting power and use it on a different candidate. This is essentially what **Instant Runoff Voting** is all about, as we'll learn in the next chapter.

6. This is a slightly different question than the previous one. Instead of giving the supporters of the last place candidate a second chance to vote, what if their votes just disappeared?

Obviously, Capano would still win. But it turns out the victory appears larger, in terms of percentage. The important thing to know is that percentages change when their *denominator* (what they are "out of") changes.

- If Net's votes were not counted, then the total number of votes would be $1,840 + 1,422 = 3,262$. This is the *denominator* changing.

- Capano's new vote share:
  $\frac{1,840}{3,262} = 0.56407\ldots \approx 56.41\%$

- Russo's new vote share:
  $\frac{1,422}{3,262} = 0.43592\ldots \approx 43.59\%$

Indeed, Capano still wins, and with a majority. But the *difference in votes* between Capano and Russo did not change! It is still 1,840 votes to 1,422 votes, a difference of 418. The only thing that changed is the *denominator* (the "out of"), and whether or not we included Net's 583 votes.

Let that be another general lesson from this example. Whenever you read or hear about percentages anywhere, you should ask: "Okay, but *out of **what**, though?*" Likewise, whenever you hear about a difference in numbers, you should ask: "Okay, but how big is that *relatively speaking?*" (A difference of 418 votes is pretty big in an election of 2,000 voters, but it's minuscule in a state election with 1 million voters, for instance.)

▲

1.5 Exercises

Each chapter of this book also includes a few exercises to help you practice applying your knowledge and skills. These exercises might involve definitions, facts, and methods we've learned throughout the book so far. Sometimes, they might involve looking something up online or finding a reference. And, sometimes, they might be more open-ended, asking you to share an opinion, create an example, or find evidence to support a proposed argument.

1. Suppose a college's senior class, with 500 eligible voters, holds an election for their Student Government President using Plurality Voting.

   (a) Assume there are just two candidates. How many votes does a candidate need to guarantee a victory?

   (b) Next, assume there are three candidates. If one of them receives 125 votes, is that enough to *guarantee* a victory, without knowing anything else about the election results or voter turnout?

   To explore this, make up two example election results for the three candidates (A, B, C). In one example, make Candidate A win overall with 125 votes; in another example, make Candidate A get 125 votes but lose overall.

   (c) Still assuming three candidates, how many votes would a candidate need to *guarantee* victory, regardless of overall voter turnout?

   Then, assuming 100% voter turnout, what is the *smallest number* of votes that a candidate could receive and still win?

(d) What if there were four candidates? Or five? Or six? In each case, how many votes would *guarantee* victory, regardless of voter turnout? And then, assuming 100% turnout, what's the minimum number of votes that *could* be a winning total?

2. This question concerns the meaning of the word **plurality**. See Figure 1 for a screenshot of some election results from the Connecticut Secretary of State's website. At the bottom, notice the statement: *"Plurality: 68."*

Figure 1: Results from the January 9, 2018 Special Election for Connecticut's 15th Assembly District. Source: Connecticut Secretary of State's website [7]

SPECIAL ELECTION – JANUARY 9, 2018
VOTE FOR STATE REPRESENTATIVE
15TH ASSEMBLY DISTRICT

15TH ASSEMBLY DISTRICT	BOBBY GIBSON, JR. (D)	JOSEPH M. SUGGS, JR. (PC)
BLOOMFIELD	881	748
WINDSOR	41	106
TOTALS:	922	854

PC = Petitioning Candidate

Plurality: 68

Using the numbers in the table, identify where the value of 68 came from. Is this how we have used the word *"plurality"*? If not, what word or phrase might more accurately describe what that number 68 represents?

Then, consult a dictionary and see whether their definition includes multiple senses of the word. Do they include the way we defined *"plurality"* in Definition 1.2? Do they include the way it's used in Figure 1?

3. This question also concerns the election results displayed in Figure 1, as well as Connecticut state law's stipulations for when a close election triggers an automatic recount to verify the results. According to *Ballotpedia*, a recount is mandated "when the margin is less than 0.5% of total votes cast for office but not more than 2,000 votes, or fewer than 20 votes." [8]

   (a) What was the margin of victory in the election whose results are shown in Figure 1? What is that margin as a percentage? Does this mandate an automatic recount or not?

   (b) With the given number of total voters, what would the margin of victory (# of votes) have to be to trigger an automatic recount?

   Then, visit the website where those election results are shared [7] and notice that, in fact, 1,787 voters cast a ballot in the election. Does that change your answer to this question, or the previous one, at all? And why do you think the *total # of voters* you were working with in part (a) was not 1,787?

   (c) Let's set the specific Gibson vs. Suggs election aside and consider Connecticut's recount stipulations more generally.

   How many total votes would have to be cast for a margin of victory of exactly 2,000 votes to amount to 0.5%? Then, make up an example of election results where Candidate A defeats Candidate B by more than 2,000 votes but less than 0.5%.

   (d) Similarly, how many total votes would have to be cast for a margin of 20 votes to amount to 0.5%? Then, make up an example where A beats B by fewer than 20 votes but greater than 0.5%.

(e) Finally, reflecting on your answers throughout this question, speculate on why Connecticut has these rules regarding recounts. What would be different if they only had the *"less than 0.5%"* condition?

4. Identify a situation in your life, *other than voting in a political election*, where you have used the Plurality Voting method. It doesn't matter whether you called it that or recognized it as such at the time, but you can probably think of a situation where this method (or something essentially like it) was used to make a group decision.

If possible, reflect on that situation. Why was Plurality Voting used? Look at the Pros & Cons list of Section 1.3 and discuss whether any of those ideas now make more sense, when considered in the specific context you have in mind.

5. Find an example of the results of a political election with three or more viable candidates where the winner received a mere plurality, not a majority, of all votes cast. (Consider using some of the links shared throughout this chapter or consult your local, state, or federal government website.)

Then, consider what might happen if the voters who supported the *last place* candidate were allowed to recast their ballot for someone else. Could that change the results of the election? Could it make a different candidate win instead? And if those voters' ballots were simply removed entirely, how would that change the original margin of victory as a *percentage?*

(For inspiration and guidance, check out Practice Problem 1.1 where we did something similar.)

Chapter 2:
Runoff Methods, including Instant Runoff ("Ranked Choice Voting")

T HE goal of this chapter is to introduce you to two related methods that both involve a **runoff** of some kind. This refers to a voting method where there are *multiple rounds of voting*, with some stipulations for when the process concludes. One of those methods is **Instant Runoff Voting**, which has gained the name **Ranked Choice Voting** in the news media in recent years. We will discuss where this method is currently used, as well as some efforts underway in various places to adopt this voting method.

2.1   Definitions and Examples

Before defining some voting methods, we must introduce an important idea about a voter's **ballot**. To motivate this idea, let's briefly return to our example of nine friends voting on dinner.

Example 2.1. Remember: of the nine friends, four of them named Indian food as their top choice, three chose Pizza, and two chose Thai food. Therefore, the winner under the Plurality Voting method would have been Indian.

But let's say that the friends talked amongst themselves and no-ticed that the winner was merely a **plurality** and not a **majority**. (Moreover, perhaps they realize that the five voters who didn't pick Indian food really *hate* that cuisine, and would have ranked it last out of all the options.) Perhaps they come to a consensus that, al-though Indian food has *the most votes*, it does not truly reflect the *overall preferences of the group*.

What could they do in that situation? Well, they might look at Thai food, which received *the fewest votes*, and say, "We cannot choose Thai food because it has so few people naming it as their top choice. So, let's ask those two friends who voted for it: **What would your second choice be?**"

Let's say, hypothetically, that those two friends do really dislike Indian food, and they name Pizza as their second favorite. In that case, we would have $3 + 2 = 5$ voters opting for Pizza, with the other four friends still opting for Indian food. That would make Pizza the overall winner!

There are two things about this example we want to point out because they directly motivate the definitions we are going to state shortly.

1. The voting method that we just used in that example worked like this: There was *no majority winner* in the first round of voting, so the candidate that was in last place in that round (Thai) was eliminated. The voters who chose that option did not have their voting power disappear; rather, *their voting power gets transferred down to their next choice*. These are precisely the rules of **Instant Runoff Voting**!

2. In that hypothetical example, we imagined that the friends held the first round of voting by just asking for everyone's

top choice. Only after they noticed there was no majority winner did they decide to follow up and ask for some voters' second place choices. However, couldn't they have asked for that information up front, on the first round of voting?

That is, instead of just writing "Pizza" as their top choice, *each voter could rank the candidates, in their order of pref- erence.* For example, a friend might write, "1st: Thai, 2nd: Pizza, 3rd: Indian." From now on, we will call something like this a **ranked ballot** or **preference list ballot**.

2.1.1 Ranked Ballots

Definition 2.1: Ranked Ballot, or Preference List Ballot.
*A **ranked ballot** is how a voter submits a rank ordering of the can- didates, based on that voter's individual preferences. Ties in the ranking are not allowed.*

*If a voter's ballot includes all of the candidates in their ranking, we will call it a **full ranked ballot**. (A voter may be allowed to submit a ranked ballot that is not full.)*

Example 2.2 (Ranked Ballots in the Real World)**.** In practice, a ranked ballot may have a list of all the candidates going down the side, and then a grid of bubbles, with columns indicating "1st choice, 2nd choice, 3rd choice, . . . " Or, it might have a list of names with empty boxes next to each one, and you need to write "1, 2, 3, . . . " to indicate your order of preference. We have included an example of each one below.

1. We mentioned before that Maine started to use Instant Runoff (Ranked Choice) Voting in 2018 for some statewide races. Before it was first used on election day, Maine's Secretary of State published information on their website, including

a sample ballot for an upcoming race. Figure 1 has a sample ballot for the Democratic Primary Election for Governor. (The primary was held June 12, 2018, and the general election was held November 6, 2018.) Notice that it uses the **grid of bubbles** method.

Figure 1: Example of a **ranked ballot** from the Democratic Primary for the election for Governor of Maine in 2018. Source: *Portland Press Herald* [9]

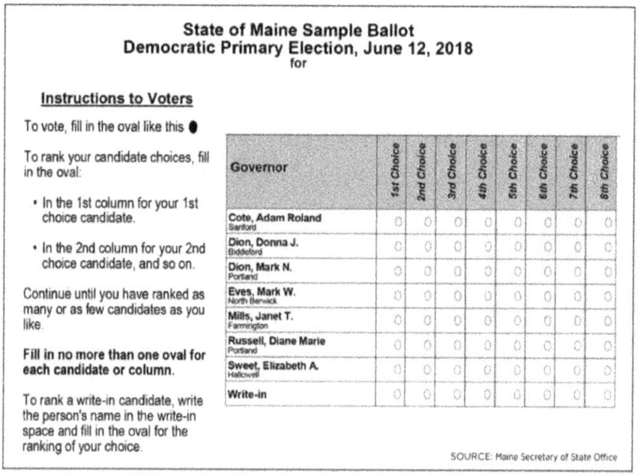

2. Australia has used Instant Runoff Voting for their House of Representatives since 1919. Figure 2 has a sample ballot from the Australian state of Victoria. Notice that it uses the **number the boxes** method.

It is important to mention a few rules about ranked ballots. These ideas are important not only in real world usage, but also in terms of the mathematical techniques we will learn and apply later.

1. **No ties are allowed in the rankings.** We stated this in the definition above, but it bears repeating. When we talk about

Figure 2: Example of a **ranked ballot** from an election for the Australian House of Representatives. Source: Wikipedia [10]

ranked ballots, we specifically mean that voters are ranking the candidates in an order of their choosing, but they *cannot* say something like, "These two are my favorites so I'm ranking them both first."

Look to the example ballots above to consider this in real world usage. In the *grid of bubbles* ballot, this means a voter cannot put two bubbles in the same row (that would be ranking a candidate in two different positions), nor can they put two bubbles in the same column (that would be ranking two different candidates at the same position). In the *number the boxes* ballot, this means a voter cannot write the same number in two different boxes, nor can they write two numbers in the same box.

These are important issues, because some opponents of voting methods like Instant Runoff Voting argue that ranking the candidates will cause more **voter error**, leading to ballots that are thrown out because they don't make sense. There are some studies about this, and they seem to indicate that voter error is not necessarily more prevalent in Instant Runoff Voting, as compared to Plurality Voting [11, 12, 13, 14]. However, in the examples in this book, we will focus on the mathematical issues of voting methods and will assume that every voter correctly fills out their ballot with no ties in their ranking.

2. **Voters do not have to rank *all* candidates, but it will often be convenient to assume that everyone does.** It is possible for a voter to have very specific preferences and not want to apply a ranking to all of the candidates. We will allow this to happen (as most real world elections do) in some examples. For instance, notice that the Maine example ballot above says: "Continue until you have ranked as many or as few candidates as you like."

In our example of friends ordering dinner, perhaps I really want Indian food but I have no strong feelings between the other two options (Pizza and Thai). So, even if I were asked to rank the three options in order, I may choose to write, "1st: Indian" and nothing else. On the *grid of bubbles* ballot, that would be equivalent to filling in only the first column and leaving the rest blank. On the *number the boxes* ballot, that would be equivalent to only writing "1" in a single box and leaving the rest blank.

There is nothing against doing this. My vote would not "count less" in any way. The only thing that could happen is that *my voting power disappears if my ballot gets "used up."* (This is what the term **"exhausted ballot"** means.) For example, if I only ranked Indian food first, and it turns out that Indian gets eliminated (like we saw Thai get eliminated in the hypothetical example above), then I would have no second choice to fall back on and I would not be included in the second round of voting. This may make some calculations in the second round (or later rounds) more complicated, but it is not against the rules at all. Moreover, this is a perfectly valid opinion for a voter to have and express on their ballot.

In the examples in this book, it will sometimes be helpful to assume that every voter actually ranks all the candidates. This may make some calculations or logical arguments much easier to understand. We will point out when we are making such an assumption by saying something like: "Assume that every voter submits a *full* **ranked ballot**."

2.1.2 Instant Runoff (Ranked Choice)

We are now ready to state the rules of Instant Runoff Voting. Afterwards, we will apply the method to several examples so you see how it works. You may notice that, from now on, we will always be assuming there are *at least three candidates* in any election. If there were only two candidates, Majority Rules would be the obvious method to use. And, in fact, for all of these methods we will learn about, if we apply their rules to a situation with just two candidates, they would work exactly the same as Majority Rules anyway!

Definition 2.2: Instant Runoff Voting.
Assume there are ≥ 3 candidates and voters submit ranked ballots (with no ties). Several rounds will be conducted, using the voters' ranked ballots. The main idea is to use the "1st place rankings" and look for a majority winner.

1. *Look at the top of every voter's ranking. If some candidate has a **majority** of those 1st place votes, that candidate is declared the winner.*

2. *If there is no majority winner, identify the candidate who has the fewest 1st place votes in this round. That candidate is eliminated from the running. (Note: If there are two or more candidates tied for the fewest votes, they are all eliminated in this step.)*

3. *Take every voter's ranking and remove that eliminated candidate. If this makes a ballot **"exhausted"** (because the voter submitted a ranked ballot that is not full), that voter is removed from the process, as well.*

4. *Now, repeat the same steps. Look at the top of every voter's ranking. If a candidate has a majority of those 1st place votes (not counting any voters with exhausted ballots), they are the winner. Otherwise, identify the candidate with the fewest votes and eliminate them from the process.*

5. *Repeat these steps over and over. At some point, a candidate must have a majority. At the worst, you may have to narrow the field all the way down to two candidates (at which point it becomes Majority Rules). Or, the process may already stop in an earlier round.*

Let's apply this method (which we will call **IRV**, for short) to the example of friends voting on a cuisine. This will also introduce some important notation we will use from now on: a **table of ranked ballots**.

Example 2.3. Assume we have nine voters (Adam through Ivan) and three candidates: Indian (I), Pizza (P), and Thai (T). We ask the voters to submit a **ranked ballot** (no ties), and they each happen to submit a **full** ranked ballot (they rank all three candidates). Let's say these are the results:

- Adam: $I > T > P$
- Beatrice: $P > T > I$
- Chandler: $T > P > I$
- Deepak: $I > T > P$
- Eve: $I > T > P$

- Frankie: $T > P > I$
- Gabi: $P > T > I$
- Hector: $I > T > P$
- Ivan: $P > T > I$

We hope you know what we mean: Adam's ballot means Indian is ranked 1st, Thai is 2nd, and Pizza is 3rd. We will often use the ">" and "<" symbols to indicate rankings.

Doesn't it seem a little annoying to apply the IRV method to this list of names and rankings? Isn't it hard to count the 1st place votes easily? And after that, wouldn't it be challenging to see whether to use a voter's 1st or 2nd place ranking? We hope that you agree there must be a better way to *organize the information provided by the voters' ranked ballots.*

For instance, notice that Adam, Deepak, Eve, and Hector all supplied the same ranking: $I > T > P$. Moreover, three voters supplied the same $P > T > I$ ranking: Beatrice, Gabi, and Ivan. And finally, the remaining two voters (Chandler and Frankie) both submitted the ranking $T > P > I$. Therefore, we can organize this information by recognizing that **it only matters *how many voters* submit each possible ranking**, not *who* submits each ranking.

# of voters:	4	3	2
1st choice:	I	P	T
2nd choice:	T	T	P
3rd choice:	P	I	I

The table on the left is how we will organize the information provided by the voters on their ranked ballots; we call this a **table of ranked ballots**.

- Each column corresponds to one possible ranking of the candidates. The column heading tells us how many voters submitted that ranking.

  For instance, the first column tells us four voters submitted the ranking $I > T > P$. (It doesn't tell us *who* voted that way, just *how many*.)

- We will look to the rows when we apply the IRV method to examples like this. Specifically, we will look to the *top row*, because that corresponds to the voters' 1st place votes. In subsequent rounds of the process, we may cross out some entries in the table or just rewrite the table.

Now, let's apply the rules of IRV.

1. *Look at the top of every voter's ranking.* We see that I gets 4 votes, P gets 3 votes, and T gets 2 votes. That's a total of $4+3+2 = 9$ voters, and the highest proportion is $\frac{4}{9} \approx 44.4\%$, which is not a majority.

2. *Eliminate the candidate with the fewest 1st place votes.* In this case: T.

3. *Rewrite ballots without that candidate.* We can do this in two ways. We could entirely rewrite the table of ranked ballots as if T were no longer involved. Anywhere T appears, we cut it

out; any candidates ranked below that point will be bumped up. The result is on the left below.

Or, we could do cross out T wherever it appears and rely on our eyes to carefully "skip over" any rows that have something crossed out. The result is on the right below.

# of voters:	4	3	2
1st choice:	I	P	P
2nd choice:	P	I	I

# of voters:	4	3	2
1st choice:	I	P	$\not{T}$
2nd choice:	$\not{T}$	$\not{T}$	P
3rd choice:	P	I	I

4. *Continue by repeating those steps.* Look at just the 1st place votes.

   In the table on the left, we can easily see that P gets $3+2 = 5$, which is a majority because $\frac{5}{9} \approx 55.6\% > 50\%$.

   In the table on the right, it may be a little harder to see, but the results are the same. In the far right column, we just have to look to the second row to see those voters' "1st place vote" in this round, because their true top choice has been eliminated. We can see that P gets $3 + 2 = 5$ votes.

   The analysis of this example wasn't too involved because we only had three candidates and three columns. In larger examples, especially with more candidates, it may become tricky to keep track of all of the required information for each round of the process. So, *we recommend rewriting the table of ballots in each step.*

5. *Conclusion: Pizza (P) is the winner.*

Example 2.4. Let's consider a larger example of an election, with four candidates and eighteen voters. The table of ranked ballots is shown below. Just like in this example, we will often use generic letters for candidates: A, B, C, ...

This time, let's allow the voters to not rank *all* four candidates. The "-" symbol in the table indicates a *ranked ballot that is not full*. For example, the far right column in the table shows one voter who only indicated that D is their top choice, with no further preferences among the rest. For another example, the two voters represented by the fourth column all ranked C 1st and D 2nd, but they are saying they're indifferent if the election comes down to A vs. B.

# of voters	6	4	2	2	1	2	1
1st	A	B	C	C	C	D	D
2nd	C	D	A	D	B	B	-
3rd	D	C	B	-	-	A	-
4th	-	-	D	-	-	C	-

1. Tally the 1st place votes: A has 6, B has 4, C has $2+2+1 = 5$, and D has $2 + 1 = 3$. The total number of votes is therefore $6+4+5+3 = 18$, so a majority requires 10 votes, which no one has. (Remember: $\frac{9}{18} = \frac{1}{2} = 50\%$ is not quite a majority; you need *more* than half.)

2. The candidate with the fewest 1st place votes is D. Let's rewrite the table of ranked ballots as if D were no longer in the running:

# of voters	6	4	2	2	1	2	1
1st	A	B	C	C	C	B	-
2nd	C	C	A	-	B	A	-
3rd	-	-	B	-	-	C	-

We have continued to use "-" to indicate a non-full ranked ballot. Notice that the voter represented by the far right column is essentially *no longer counted*; their ballot is **exhausted**. So, in the next step, when we look for a majority of 1st place votes, keep in mind that there are only *17 voters* now, not the original 18.

(Here's another way to think about that *"voting power disappearing."* The voter's ranking indicated that, beyond *D* as their top choice, they don't have any preferences among the other three. It's like the voter is saying: "If *D* doesn't win, I don't care what happens; the rest of y'all can decide.")

3. Tally the 1st place votes: *A* has 6, *B* has $4+2 = 6$, and *C* has $2 + 2 + 1 = 5$. A majority would require 9 out of 17, which no candidate has, so we eliminate *C* next. (It does not matter that *A* and *B* are "tied for 1st" now; we're only looking for who has the *fewest* 1st place votes.)

4. Now it's just a Majority Rules contest between *A* and *B*:

# of voters	6	4	2	2	1	2	1
1st	*A*	*B*	*A*	-	*B*	*B*	-
2nd	-	-	*B*	-	-	*A*	-

We see that *A* gets $6+2 = 8$ votes, while *B* gets $4+1+2 = 7$. This means *A* must win, but is it a **majority** victory, as this method is supposed to find? It is, as long as we remember to exclude any exhausted ballots!

The two voters in the fourth column have now disappeared from the process, just like the one voter in the far right column. We wish to emphasize that *this is a **feature** of the method, not a bug!* Those three voters used their ballots to

35

express preferences about only candidates C and D, who happened to not advance to the final round. Effectively, those voters were saying, "If it's a race between just A and B, we have no preference; y'all can decide." Perhaps those voters would have just stayed home and not voted, if A and B were the only candidates in the race. So, in fact, this is an arguable *benefit* to allowing ranked ballots, and non-full rankings, at that: this might encourage voters who would otherwise sit out an election to instead participate and express their preferences.

Whatever their reasons, those three voters are excluded from the process now, and we find that A wins with a majority: $\frac{8}{15} \approx 53.3\%$. That is, A has a majority of the **continuing ballots** ("not exhausted"). You may notice that this is *not* a majority of *all ballots* originally cast: there were 18 voters at first, and $\frac{8}{18} \approx 44.4\%$. However, this does not cast any doubt on A's victory, and it does not mean it is merely a plurality win. Rather, A's victory truly is a majority. It's just that some of the electorate exercised their right to express no preference about A vs. B, so it's as if they had not even participated in the first place. Their voting power is not "less than" the 15 remaining voters, and their voices did not get ignored; they just got "used up" in the process.

By the way, notice that some other voters submitted ranked ballots that were not full, and they didn't end up as exhausted ballots. For example, the one voter in column five, who ranked only $C > B$, had their ballot continue throughout every round. This is because at least one of the candidates on their ballot (namely, B) made it all the way to the final round. In general, voters would have no way of knowing how everything will play out, so it's unfair to

36

say that some ballots ended up being "worth less" than others or anything like that. This is simply how the Instant Runoff Voting method works: it looks for a majority winner, but the threshold for achieving a majority may decrease with each round as ballots get exhausted.

One final observation from that example: candidate A would have been the winner under Plurality Voting, too! While it's true that *the choice of voting method determines the winner just as much the voters' ballots do*, we don't want you to assume that the Instant Runoff method will *always* disagree with the Plurality method.

Let's see just one more example to remind you of an important stipulation in the IRV rules and to preview an idea we'll explore in much more detail later.

Example 2.5. Consider an election with three candidates and twenty-one voters. Their preferences are shown in the table of ranked ballots below.

# of voters	9	6	5	1
1st	A	B	C	C
2nd	B	A	B	A
3rd	C	C	A	B

1. Tally the 1st place votes: A has 9, B has 6, and C has $5+1 = 6$.

   A majority requires 11 out of 21 votes, so no candidate wins yet. However, it is clear that A would be the winner under the Plurality Voting method.

2. Who to eliminate in this round? Both B and C are tied for the *fewest 1st place votes*, and the IRV rules have no standard procedure for breaking such a tie. Instead, remember that the rules (as stated in Definition 2.2) say to *eliminate all of them*.

In this case, that means B and C are eliminated and only A advances to the second round. In other words, A wins outright because they're the only candidate left!

(We will, in fact, revisit this example later in the book to demonstrate a particular property of the IRV method that could be construed as a flaw. If you are curious right now, though, here's something you can try: In this example, what if the one voter represented by the far right column *changed their ballot* to be $A > C > B$ instead? Apply the IRV method to determine the winner. You should find that it's B, and you should think about why that might be a strange phenomenon, considering that A won in the original example ...)

One final comment about the Instant Runoff Voting method before we move on to a similar method. It is called **"Instant" Runoff** because the runoffs (the separate "rounds" of voting) are all conducted instantly using the voters' ranked ballots that they submitted all at once. In other words, we do **not** have to do what we alluded to in Example 2.1, where the friends noticed there was no majority winner so they had to *ask some voters to submit more information.* In real world usage, this is very important; indeed, supporters of this method often mention the fact that it can be costly and time-consuming to conduct entirely separate runoff elections (like primaries before a general election).

2.1.3 Top Two Runoff

The next voting method we will learn is similar to the IRV method in that it involves a runoff, a separate round of voting. However, the main difference is that *there are only two rounds.* Using the IRV rules, we eliminate candidates one by one until one of them has a majority of 1st place votes. In the **Top Two Runoff** method, the second round narrows the field down to just two candidates right away. This is sometimes called **Plurality Runoff** or **Two Round Runoff**, because it essentially uses two rounds of plurality voting. We will use the name **Top Two Runoff** to remind you that the second round narrows the field to the top two candidates.

Definition 2.3: Top Two Runoff Voting.
Assume there are ≥ 3 *candidates and voters submit ranked ballots (with no ties). Two rounds will be conducted, using the voters' ranked ballots. The main idea is to use the "1st place rankings" and narrow down to the top two.*

1. *Look at the top of every voter's ranking. If some candidate has a **majority** of those 1st place votes, that candidate is declared the winner.*

2. *If there is no majority winner, identify the **top two** candidates with **the most 1st place votes** in this round. Everyone else is eliminated from the running. (Note: If there are two or more candidates tied for the top two places, they are all included in the runoff round. In other words, the field narrows down to the top two plus any ties.)*

3. *Take every voter's ranking and remove all eliminated candidates. If this makes a ballot "exhausted," that voter is removed from the process, too.*

4. *Now, look at the modified table of ranked ballots and look at just the 1st place ranks. Whichever remaining candidate has the most 1st place votes in this round, they are declared the winner. (Note: If there are only two candidates, this will be a majority, but it might **not** be if three or more candidates advanced to the runoff.)*

Let's confirm something that you may be wondering because these rules are so similar to the rules for IRV:

> If there are exactly three candidates, then Instant Runoff Voting and Top Two Runoff Voting are identical!

That is, in the first round, IRV says "Eliminate one of the candidates" and Top Two Runoff says "Narrow down to the top two candidates." But, when there are only three candidates to start with, that's the same thing! In other words, if we want to see an interesting example, we better have four candidates. Let's revisit an example from earlier:

Example 2.6. Consider the election from Example 2.4 with four candidates and eighteen voters. Their preferences are shown in the table below.

# of voters	6	4	2	2	1	2	1
1st	A	B	C	C	C	D	D
2nd	C	D	A	D	B	B	-
3rd	D	C	B	-	-	A	-
4th	-	-	D	-	-	C	-

Recall that, in this election, A was the winner using Plurality Voting and B was the winner using Instant Runoff Voting. Let's see what happens if we apply the Top Two Runoff Voting method.

1. Tally the 1st place votes: A has 6, B has 4, C has 5, D has 3. Nobody has a majority (which requires 10 out of 18), so only the top two vote-getters will advance to the final round: A (6) and C (5) advance, while B (4) and D (3) are eliminated.

2. Here's the table of ranked ballots with B and D eliminated. Notice that one ballot is exhausted, so the calculations are now based on 17 voters.

# of voters	6	4	2	2	1	2	1
1st	A	C	C	C	C	A	-
2nd	C	-	A	-	-	C	-

However, there isn't much calculating to do: A gets $6+2=8$ votes and C gets $4+2+2+1=9$, so C wins 9-8.

Notice that this is a majority of **continuing ballots** ($\frac{9}{17} \approx$ 52.9%), even though it's not technically a majority of all original ballots ($\frac{9}{18} = 50\%$).

Finally, we note that this example demonstrates that Instant Runoff and Top Two Runoff are not fundamentally identical: here we have an instance where IRV led to a different winner (A) than the Top Two Runoff method (C).

2.2 Real World Usage

After Plurality Voting, it appears that Instant Runoff Voting is the next most popular method used worldwide. Some countries use the IRV method for certain elections, and some even use variations on the method for elections where there are *multiple winners*. In some countries, they use something like IRV or Top Two Runoff but they actually hold separate elections for the runoff rounds, instead of doing things "instantly" with ranked ballots.

Regardless of actual political races, though, we think you are somewhat familiar with the mechanics of these voting methods just from having to make group decisions in your life. Have you ever had a large group of people, like a class of students or an athletic team, choosing one option from a big list? Perhaps you had a group discussion that narrowed the options down to the collective "top two," and then you raised your hands to choose from between those two. That's essentially an informal version of the Top Two Runoff method, just without the ranked ballots and the instantaneous runoff.

2.2.1 Variations on IRV for Multiple Winners

In this book, we are focused on elections with *one winner*, like a race for a Senate seat. However, plenty of political races involve *multiple winners*.

For instance, the City Council in Cambridge, Massachusetts has nine members, and elections are held every two years for all of those seats. Paper ballots use the *grid of bubbles* method to allow voters to rank the candidates. These rankings are then used to decide all nine seats on the council. The technical term for the method that Cambridge, MA uses is **Single Transferable Vote (STV)**, and the website of the city's Election Commission has some information about how the calculations are done. [15, 16]

We may look at voting methods like this in a follow-up to this book. For now, we just wanted to mention this distinction between *single winner* and *multi-winner* elections. Moreover, you may hear terms like **STV** used if you read about modern voting methods, so you should be aware that they're referring to something that is very similar to IRV but not quite the same.

2.2.2 Real World Examples of Elections

France: Top Two Runoff with a Separate Election

The President of the French Republic is elected every five years by a method that is fundamentally equivalent to the **Top Two Runoff** method we have learned. However, the *logistics* of the election make it sightly different from that mathematical method, as we have described it.

In the definition and examples earlier, you saw that the Top Two Runoff method involves each voter submitting a ranked ballot where they can list the candidates in their order of preference. The method then says to look at just the 1st place votes: if some candidate has a *majority*, they are declared the winner. Otherwise, only the top two candidates with the *largest pluralities (most votes)* advance to the runoff round. Now, imagine you are a voter and your top-ranked candidate did *not* advance to the runoff round. Your voting power would not disappear: it would get transferred down to *whichever of the top two candidates is higher on your ranking.* (Remember: we are allowing voters to submit rankings that are not *full*, so it is possible that your vote disappears, but only because you did not list all candidates on your ranking.) So, the initial round and the runoff round can be conducted all at once, using the voters' ranked ballots.

In actual practice, voting for the President of France is conducted via *two separate elections* using *vote-for-one ballots*, not ranked ballots. The initial round is held sometime in April and is conducted like a Plurality Voting election: voters must choose *one* of all the candidates. If anyone receives a majority, they automatically win. If no one has a majority, then a separate election is held two weeks later: it's a Majority Rules vote between just the top two vote-getters from that initial round. Voters come back and select one of the two contenders, and the one with the most votes wins.

Notice how this is different from the method we described earlier because there are essentially *two separate elections with Plurality Voting ballots* instead of one election with ranked ballots. Theoretically, this could be done all at once with ranked ballots, instead of asking voters to return to the polls two weeks later. So, think about what the benefits might be of having a separate second round. Perhaps there are so many candidates that voters will have trouble ranking *all* of them, and asking them to just choose one for now makes it *easier to decide*, and therefore increases voter turnout. And, perhaps a voter would want to essentially change their ranking after seeing how the rest of the electorate voted. However, there can be close races in the first round, where slightly different results there could have led to vastly different final results.

Example 2.7. Let's look at France's 2017 Presidential election. The first round, held on April 23, had at least eleven candidates. The top four vote-getters each received around 20% while no other candidate earned more than 6.5%. [17]

1. Emmanuel Macron, 8,656,346 votes, 24.01%

2. Marine Le Pen, 7,678,491 votes, 21.30%

3. François Fillon, 7,212,995 votes, 20.01%

4. Jean-Luc Mélenchon, 7,059,951 votes, 19.58%

It was a narrow race at the top, and it's easy to imagine the voters' strategic decisions changing slightly and causing the top two candidates to be different. In any event, Macron and Le Pen advanced to the second round, because no one had a majority of the votes. (Interestingly enough, France has held Presidential elections in this manner since 1965, and there has *never been a majority winner in the first round;* it has gone to a runoff round every time!)

That second round was held two weeks later on May 7, 2017. Although their vote shares were close in the first round, Macron won easily over Le Pen in the second round, with almost twice as many votes:

1. Emmanuel Macron, 20,743,128 votes, 66.10%

2. Marine Le Pen, 10,638,475 votes, 33.90%

Presumably, everyone that voted for Macron in the first round also voted for him in the second round, and the same goes for Le Pen. (However, because they were actually two separate elections, we cannot be 100% certain of this, even though it seems like a reasonable assumption.) So, notice that Le Pen gained approximately 3 million votes in the second round, while Macron gained about 12 million votes. You might think that this means most of the first-round voters switched their votes to Macron for the second round, and only some switched their votes to Le Pen. But it's even more complicated than that, because *about 5 million fewer people voted in the second round than the first!* [17]

All of this is to say that this method is superficially similar to the Top Two Runoff method we have described in this chapter, but not exactly the same. The real world logistics of these French elections may have significant effects on the behavior of the voters, and it is not fair to say that it is *"identical"* to the mathematical method we have learned. In short, voting in the real world can be messy and unpredictable. But that doesn't make our mathematical study of voting unhelpful or unrealistic. It's just a way to help us better understand the messiness of the real world.

Maine: Instant Runoff in 2018

We have mentioned that Maine passed a ballot referendum in 2016 calling for Instant Runoff Voting to be used in future elections. After a prolonged political battle involving public opinion, petitions for signatures, and court proceedings, the first time the method was actually used for a statewide election in the U.S. was in November

2018. In Maine's 2nd Congressional District, the mechanics of IRV led to an overall winner that was *not the plurality winner*.

The incumbent in that district, Republican Bruce Poliquin, faced three opponents in the general election: Democrat Jared Golden, and two Independents, Will Hoar and Tiffany Bond. If you consult *Ballotpedia* for the results, reproduced here in Figure 3, you may be a little confused. It would appear that the two Independents got *literally no votes*, while the other two candidates were almost evenly split, with Golden getting a very slight majority.

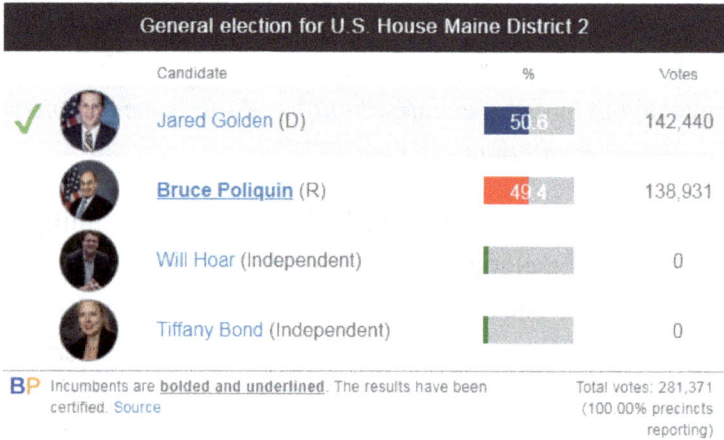

General election for U.S. House Maine District 2		
Candidate	%	Votes
✓ Jared Golden (D)	50.6	142,440
Bruce Poliquin (R)	49.4	138,931
Will Hoar (Independent)		0
Tiffany Bond (Independent)		0

BP Incumbents are **bolded and underlined**. The results have been certified. Source

Total votes: 281,371
(100.00% precincts reporting)

Figure 3: The final results of the general election for Maine's District 2 in the U.S. House of Representatives in 2018. Source: Screenshot of *Ballotpedia* [18]

In fact, this is what resulted *after the IRV procedure was followed*. That is, it's not the list of 1st place votes from the first round; those results are shown in Figure 4. Notice that Poliquin actually had a slim plurality. Under the usual Plurality Voting rules, that would be the end of it: Poliquin would win with 46.33% of the vote, more than any other candidate.

U.S. House, Maine District 2 General first round, 2018			[hide]
Party	Candidate	Vote %	Votes
■ Republican	Bruce Poliquin *Incumbent*	46.33%	134,184
■ Democratic	Jared Golden	45.58%	132,013
Independent	Tiffany Bond	5.71%	16,552
Independent	Will Hoar	02.37%	6,875
Total Votes			**289,624**

Source: *Maine Secretary of State*, "Tabulations for Elections held in 2018," accessed January 7, 2019 ⏎

Figure 4: Results of the first round of the general election for Maine's District 2 in the U.S. House of Reps. in 2018. Source: Screenshot of *Ballotpedia* [18]

Instead, the IRV rules say that we should eliminate the candidate with the fewest votes: Will Hoar with only 2.37%. However, even if *literally every single one of those voters* ranked Tiffany Bond as their 2nd choice, that would still only give her approximately 8% of the vote ($5.71 + 2.37 = 8.08$); so, it is mathematically certain that Hoar will be eliminated in the first round and then Bond will be eliminated in the second round. In practice, this means that all of this can happen in one step: on every ballot that listed *either* Bond or Hoar as the top choice, look down that ballot to find whether Poliquin or Golden is ranked higher. (In other words, if somebody ranked Bond 1st and Hoar 2nd, we would be looking at their 3rd place choice, if they listed one.)

This will increase the values that you see for Poliquin and Golden in the *Votes* column in Figure 4 above. Indeed, the final values are what you saw first in Figure 3. Evidently, Poliquin gained 4,747 votes in the process ($138,931 - 134,184 = 4,747$) while Golden gained 10,427 votes ($142,440 - 132,013$). This shows that, of all the voters who ranked one of the Independent candidates 1st, *many more of those voters ranked the Democrat higher than the Republican* on the rest of their ranking. This is what turned

Poliquin's slim *plurality* of votes in the first round into Golden's slim *majority* in the runoff round.

Before leaving this example, we should point out something else that is interesting to observe and may be a point of discussion for those advocating for or against IRV in the real world. It involves the fundamental idea that IRV is meant to modify Plurality Voting while also *seeking a majority winner*. The main question becomes: a majority of *what?*

Now, certainly, Golden's gain of 10,000+ votes in the runoff was much more substantial than Poliquin's gain of just about 5,000 votes. However, it's worth contemplating the fact that *there were 8,253 exhausted ballots* in the runoff. Where does that number come from? Well, there were $16,552 + 6,875 = 23,427$ voters who ranked one of the Independents first; however, the major party candidates only gained a total of $4,747 + 10,427 = 15,174$ votes in the runoff. The difference between those accounts for the exhausted ballots: there must have been $23,427 - 15,174 = 8,253$ voters who ranked an Independent first but then *did not list **either** the Democrat or Republican on their ranking at all*. (Even if just one of those candidates were listed on such a voter's ballot, that candidate would gain their vote in the runoff; and if both candidates were listed, whichever one was listed higher would gain their vote.)

So, in the end, Golden did win according to the rules of IRV. And he did have a majority of votes *that counted in the final round*: his 142,440 votes out of the 281,371 **continuing ballots** that were **not exhausted** amounts to 50.62%. However, it is worth mentioning that he did not technically have a majority of *all ballots cast*, including the ballots that were exhausted by the end: specifically, $\frac{142,440}{289,624} \approx 49.18\%$. Is this a reason to discount the IRV process, in general, because of this possibility? We think not! It is, in fact, reasonable to expect something like this to happen when a significant number of voters exercised their right to not a express a preference about some candidates (by not listing them in their ranking). Still, you should be aware that this phenomenon is possible. In fact, we previewed this exact idea back in Example 2.4 where a similar phenomenon occurred.

The table in Figure 5 from Wikipedia organizes all of this information and you can see the difference we're alluding to by comparing the columns labeled **% (gross)** (out of *all* ballots originally cast) and **% (net)** (out of all *continuing ballots / active votes* in the final runoff round).

Party	Candidate	Round 1		Transfer	Round 3		
		Votes	%		Votes	% (gross)	% (net)
Democratic	Jared Golden	132,013	45.6%	+ 10,427	**142,440**	49.18%	**50.62%**
Republican	Bruce Poliquin (incumbent)	**134,184**	46.3%	+ 4,747	138,931	47.97%	49.38%
Independent	Tiffany Bond	16,552	5.7%	- 16,552	Eliminated		
Independent	Will Hoar	6,875	2.4%	- 6,875	Eliminated		
	Total active votes	289,624	100%		281,371		100%
	Exhausted ballots	-		+8,253	8,253	2.85%	
	Total votes	289,624	100%		289,624	100%	

Maine's 2nd congressional district, 2018 general elections[21]

% (gross) = percent of all valid votes cast (without eliminating the exhausted votes)
% (net) = percent of votes cast after eliminating the exhausted votes

Figure 5: All results for the general election for Maine's District 2 in the U.S. House of Reps. in 2018. Source: Screenshot of *Wikipedia* [19]

Other Instances of IRV Around the World

We have focused on two specific elections so far, so we will conclude this section by listing some other places around the world that use IRV.

- As illustrated by the ballot in Figure 2, Australia uses IRV for elections for their House of Representatives.

- In Canada, the Province of Ontario announced in 2014 that its municipalities have the *option* of using IRV for local elections moving forward. Some cities have already started to do so, with London, Ontario being the first, in 2018. [20]

- IRV is used to elect the President of India, but not via direct vote by the citizens. Instead, citizens vote for members of Parliament, and then those members use IRV to vote for President. [21]

- Ireland uses the **Single Transferable Vote** method we mentioned earlier, but in races where there is one winner, this is identical to IRV. So, it is fair to say Ireland uses IRV to elect their President, although the term *"instant runoff"* is not commonly used there.

- In the United States, IRV is used in local elections in several cities across the country. The website for *FairVote* (an advocacy group for Ranked Choice Voting) has a list as well as an interactive map. [22]

- Some states are making moves to use IRV in statewide elections for state legislatures and federal offices, as well.

  - We just discussed Maine's use of IRV in 2018; that was the first time IRV has ever been used in a statewide election in the US.
  - Maine also used IRV in the 2020 Presidential election, where Joe Biden received a majority ($\approx 53\%$) of 1st place votes in the first round (and, hence, no runoff rounds required). [23]

– Alaska passed a ballot referendum in 2020 that enacts a hybrid form of Plurality and Instant Runoff Voting for future statewide elections. Instead of partisan primaries – where each political party holds their own primary and puts forth one candidate in the general election – there will be an *open, top four-primary*. All candidates from any party run in the same primary election, and voters use a Plurality Voting ballot to select one candidate. The top four vote-getters then advance to the general election, where Instant Runoff Voting is used, allowing the voters to rank those four candidates. [24]

The law also stipulates IRV will even be used in statewide elections that don't have such a primary, including US Presidential elections. Finally, the same ballot referendum that enacted these changes also imposes reporting requirements on campaign donations, signaling that voters see these issues – electoral rules and campaign financing – as equally important in preserving democratic processes.

- There are many other organizations and groups that use IRV, like professional societies and university student governments. [25, 26]

A perhaps surprising example is the Academy of Motion Picture Arts and Sciences: they use IRV for nominating and selecting Best Picture! [27]

2.3   Discussion of Pros and Cons

As with the previous chapter, let's briefly discuss some Pros and Cons of the voting methods we have learned in this chapter. We will focus on Instant Runoff (Ranked Choice) Voting, as opposed to Two Round Runoff, because it has garnered much more attention in the news media recently and is more frequently used in practice.

2.3.1   Pros

1. **IRV is similar enough to Plurality Voting but is based more on finding a *majority* winner.** One of the most obvious drawbacks of Plurality Voting is splitting the votes among several candidates so that the winner has a small plurality, much less than half of the votes. IRV seeks to alleviate that by redistributing the voting power of voters who chose the less popular candidates, and allowing them to voice their opinion on the candidates that did happen to do well.

2. **IRV can encourage 3rd parties and underrepresented groups to run for (and help them win) elected office.** This is meant to directly contrast with one of the Cons of Plurality Voting that we mentioned back in Section 1.3.2. With Plurality Voting, voters may be discouraged from using their voting power to support candidates outside the political establishment and the two major parties. When those candidates are unlikely to win, their supporters are sometimes accused of "wasting their vote" on a candidate that has no chance. Indeed, we described the **"spoiler effect"** in that section.

   With IRV, voters who wish to support such candidates do not need to "waste" their vote or "spoil" other candidates. A voter may comfortably rank whoever they wish 1st, and then rank someone 2nd or 3rd in the (sometimes likely) event that their top choice does not fare well in the election. The voter's voice does not disappear, and they are still able to *express their support for a certain candidate, and have that support recorded publicly.* By this, we mean something like what we saw in the example above from Maine's 2nd Congressional District, in Section 2.2.2. The roughly 8% of voters who supported an Independent candidate as their 1st choice were able to make that support clear in the election results, so that both the candidates and the general public have a rough idea of how popular those candidates were. Moreover, in doing so, these voters were not necessarily wasting their voting power. We saw that about 15,000 of them contributed their voting power to the runoff round.

All of this seems to have consequences for the demographics of candidates and winners, not just their political affiliations. A few studies have shown that IRV leads to not only more minority candidates running for races in the first place, but also more minority candidates winning seats. In particular, *women and people of color* are more likely to run, and more likely to win, at least in the studies that have been done in the few places where IRV has been used. [28, 29, 30, 31]

Obviously, more studies need to be done, but this is promising news for anyone who wishes for more diverse and equitable representation in elected offices of all kinds. A section of *FairVote*'s website has links to several studies about this important topic. [32]

3. **IRV discourages attack ads and can even encourage campaigns supporting each other.** Another downside we mentioned in Section 1.3.2 is the tendency for Plurality Voting to encourage negative campaigning. ("Don't vote for them!" vs. "Vote for me because . . . ") IRV instead allows voters to rank the candidates and know their vote will still count, even when their top choice does not win. Accordingly, candidates

need to seek broad support from the electorate and ask voters to rank them 2nd or 3rd, even when they're not their favorite.

Over the last decade, a few U.S. cities have started using IRV in elections for Mayor and other local offices. Some news coverage of those races described positive voter satisfaction with being able to rank the candidates and with how the campaigns were conducted. [33, 34, 35, 36]

As with the previous item in this list, the *FairVote* website links to several studies on this topic of **campaign civility**. [37, 38, 39, 40]

2.3.2   Ranked Ballots and the Term "Ranked Choice"

We feel obligated to mention that, as compelling as those Pros may be, many of them could be construed merely as *reasons to use **some kind of ranked ballot method** and not Plurality Voting anymore.*

You may have noticed that we often referred to the downsides of Plurality Voting when describing the advantages of IRV. This is partly because we're not assuming you have any familiarity with voting methods *other than* Plurality Voting, simply because of how common it is in the US. (If you already know something about other methods, that's great!) But it's still a fair criticism to say that, thus far, we've mostly explained why Plurality Voting is not great and why we might consider doing *something else*, whatever that *something* might be. Keep this in mind as we learn about more voting methods that use ranked ballots (in Chapters 3 and 4) and some that don't use rankings (in Chapter 5).

It is, therefore, somewhat interesting that Instant Runoff Voting has acquired the more popular moniker **Ranked Choice Voting** in the news media over the past few years. The term "Ranked Choice" sounds more general, emphasizing the fact that voters can submit their preferences by ranking the candidates. And yet, it does not convey anything else about how the method works, how it conducts multiple runoff rounds one by one, and how it does so instantly, without needing voters to cast ballots again. Doesn't **Instant Runoff Voting** accomplish all of that? Indeed, this is why we prefer that term; we have used the two interchangeably thus far, but after this chapter, we will stick to **IRV**.

2.3.3 Cons

Most of the Pros we mentioned involved the logistics of campaigns and elections, so some of the Cons discussed here will involve those things, as well. But we will also mention a strange mathematical property of IRV that opponents of the method sometimes point to as a reason to avoid it.

1. **Voter confusion and ballot errors may be more likely.**
   Some opponents of IRV cite logistical difficulties for voters and governments alike.

   Because voters in the U.S. are accustomed to choosing just one candidate, they would need to be educated about the rules of IRV. How should this be done effectively? And how should the actual ballots, and the procedures at polling locations, be updated for the new rules? This would involve some resources (time and money) and this can boil down to a debate over whether those expenditures are worth it.

   Some people even argue that there will be significantly more errors by voters: marking ballots incorrectly because of confusing instructions, or voters misunderstanding the method and not realizing they are not voting strategically. Some studies have been done in places that switched from Plurality to

IRV, and we are not aware of any findings where the rate of voter error significantly increased after such a change. However, we can understand the argument that some voters may end up not taking advantage of the voting method. If a voter just lists one candidate, and that candidate is eliminated, their voting power disappears. If that's because they really only liked that one candidate, so be it. But if that's because they fundamentally misunderstood the rules of the election, that's bad.

2. **The increased complexity of the voting method may make recounts more likely and more costly to conduct.** It's hard to determine by *how much*, but it's reasonable to anticipate that simply processing and counting paper ballots under IRV rules will be more challenging than under Plurality Voting rules.

   Using IRV rules, we may need to factor in every voter's full ranking, and we would probably need to gather all ballots in a central location to do that safely and accurately. We have an example coming up in Section 2.4 that illustrates this point.

   It's common in election laws for an automatic hand recount to be triggered when the results are "close enough", within some threshold. (See an exercise in Section 1.5 that addresses this idea.) Will such laws apply to *each round* of the IRV process? If so, in particularly close races with three or more viable candidates, there may be several manual recounts required, with significant costs and resources.

3. **It may be difficult to get IRV passed into law in some places.** Okay, this is not really a Con with the voting method itself, but it's worth mentioning in a discussion of real world

issues. Even after the voters of Maine passed that ballot referendum in 2016 to start using IRV, there was a prolonged legal battle over whether it violated the state's constitution.

4. **IRV fails the *Monotonicity Property*.** We'll address this topic in more detail in a sequel to this book about the *mathematical properties of voting methods*, but it's worth briefly mentioning now.

   Imagine a Plurality Voting election between candidates A, B, and C, where A is the winner, B is 2nd but far from A, and C is in 3rd but just barely behind B. What might happen if a few voters who supported B had voted for A instead? Wouldn't that give A *more votes,* and they would win by even more than they had before? Yes, that is exactly what would happen. But under the IRV rules, such a situation *might possibly cause A to **lose** instead!* Huh?! This may seem outrageous right now, Exercise #2 at the end of this chapter (Section 2.5) demonstrates this phenomenon.

2.4 Practice Problems with Solutions

Practice Problem 2.1. This problem will illustrate the idea that IRV ballots cannot be counted in separate "districts" and then have the results combined. (Inspiration for this problem came from an example on RangeVoting.org [41].)

Suppose we have two districts with 17 voters each, and three candidates. The ballots are summarized in the tables below.

1. Calculate the winner of District 1 using the IRV method.

2. Calculate the winner of District 2 using the IRV method.

3. Combine the ballots from the two districts into one election. Calculate the winner of that election using the IRV method. Explain what you observe.

<table>
<tr><td colspan="4">District 1:</td><td colspan="4">District 2:</td></tr>
<tr><td># of voters:</td><td>8</td><td>5</td><td>4</td><td># of voters:</td><td>8</td><td>5</td><td>4</td></tr>
<tr><td>1st choice:</td><td>A</td><td>B</td><td>C</td><td>1st choice:</td><td>C</td><td>B</td><td>A</td></tr>
<tr><td>2nd choice:</td><td>C</td><td>A</td><td>B</td><td>2nd choice:</td><td>B</td><td>A</td><td>B</td></tr>
<tr><td>3rd choice:</td><td>B</td><td>C</td><td>A</td><td>3rd choice:</td><td>A</td><td>C</td><td>C</td></tr>
</table>

▼ **Solution:**

1. In District 1, the winner using the IRV method is B.

   (a) 1st place votes: A gets 8, B gets 5, C gets 4. Thus, C is eliminated.

   (b) The 4 voters who chose C will have their voting power transferred to their 2nd choice, B. In the runoff round, this gives B $5 + 4 = 9$ votes to A's 8 votes.

2. In District 2, the winner is also B. The exact same thing happens, it's just that A gets eliminated instead of C. In the runoff round, B's $5 + 4 = 9$ votes defeat C's 8 votes.

3. Let's combine the ballots first. The result is the table below. (Notice some columns from the two tables had the same rankings, but some did not.)

Districts 1 and 2 combined:

# of voters:	8	4	12	10
1st choice:	A	A	C	B
2nd choice:	C	B	B	A
3rd choice:	B	C	A	C

Now, let's apply the IRV method:

(a) 1st place votes: A gets $8 + 4 = 12$, B gets 10, C gets 12. So, B is eliminated first.

(b) The 10 voters who supported B all ranked A 2nd. So, in the runoff round, A gets $12 + 10 = 22$ votes and defeats C's 12 votes.

This is kinda strange. In each district *individually*, B was the winner. However, with the ballots combined, not only was B *not* the winner, but they were *eliminated first!* They didn't even make it to the runoff!

This is what we meant in Section 2.3.3 when we said ballots would have to be gathered before processing. This one example shows it is possible for the results district-by-district to *not* match the overall result. ▲

Election Result: Group Ranking

We would like to use the next problem to introduce the idea of a **group ranking**, not just "the winner," that describes the results of an election. This is an idea we will use later on; the concept applies to any voting method using ranked ballots, so that includes the ones we have learned (Plurality, Instant Runoff, Top Two Runoff) as well as the ones we will learn in Chapters 3 and 4.

Definition 2.4: Group Ranking.
*Suppose an election is held using a ranked ballot method. The **group ranking** is the result of the election that identifies which candidate is in 1st place overall (the winner), which is in 2nd overall, and so on.*

1. *Apply the specified ranked ballot method to the original set of ballots to identify the winner. That candidate is 1st overall in the group ranking.*

2. *Next, remove that candidate from all the ballots, as if they were not in the running. (If this causes any ballots to be exhausted, remove them from the rest of the process.) Now, apply the specified ranked ballot method to these updated ballots to identify the "winner" there. That candidate is 2nd overall in the group ranking.*

3. *Repeat that process. Remove the candidate from all ballots, apply the ranked ballot method to identify the "winner", put that candidate next in the overall group ranking.*

The step of *removing a candidate from all the ballots* is similar to what we do in the middle of applying the IRV method. So, that's not a new idea for us. What's new here is the idea of going beyond just "the winner" of the election and recognizing that, because the voters are submitting **individual rankings** of the candidates, they can be combined to produce an overall **group ranking** of the candidates.

We must emphasize, after observing some common misunderstandings of this concept, that the group ranking is not always what you'd expect from the *"first round" results of an election*. In the problem below, pay attention to what happens when we apply the Plurality Voting method to see what we mean.

Practice Problem 2.2. Suppose an election has four candidates and fourteen voters. Their ranked ballots are summarized in the table below.

# of voters:	5	4	3	2
1st choice:	A	B	C	D
2nd choice:	C	C	D	C
3rd choice:	B	D	B	B
4th choice:	D	A	A	A

1. Summarize the results of this election by determining the **group ranking** of the candidates using the Plurality Voting method.

2. Do the same with the Instant Runoff Voting method.

3. Do the same with the Top Two Runoff Voting method.

▼ **Solution:**

1. The group ranking using Plurality Voting will be $A > C > B > D$. That is, A finished 1st, C 2nd, B 3rd, and D 4th.

   Right away, it is worth pointing out that this is *not the same as the ranking based merely on 1st place votes!* Notice that B has the second-most 1st place votes (4 votes to A's 5), but B ends up 3rd overall in the group ranking. Let's see why that happens...

   (a) Let's apply the Plurality Voting rules to the original ballots. We see A gets 5 votes, B gets 4, C gets 3, and D gets 2. Because A has the most, they are the winner. This is why A is 1st in the group ranking.

Brendan W. Sullivan

(b) Next, let's eliminate A from the ballots (table below). We will now apply Plurality Voting to *this* set of ballots.

# of voters:	5	4	3	2
1st choice:	C	B	C	D
2nd choice:	B	C	D	C
3rd choice:	D	D	B	B

Again, we only look at the 1st place votes: B gets 4, C gets $5 + 3 = 8$, D gets 2. Because C has the most, they are the "winner" here, meaning they are 2nd in the group ranking.

(c) Next, let's eliminate C from the ballots (table below). We will now apply Plurality Voting to *this* set of ballots.

# of voters:	5	4	3	2
1st choice:	B	B	D	D
2nd choice:	D	D	B	B

This is essentially a Majority Rules vote now: B gets $5+4 = 9$ votes, D gets $3+2 = 5$. So, B is the "winner" here, meaning they are 3rd in the group ranking.

(d) By default, this means D is 4th in the group ranking.

2. The group ranking using Instant Runoff Voting will be $C > D > B > A$. (Notice that A was *first* using Plurality but will be *last* using IRV!)

(a) Let's apply the IRV rules to the original ballots. Because D has the fewest 1st place votes (only 2), they are eliminated in the first round.

62

Below are the ballots with D eliminated. We must apply the IRV rules to these ballots. A gets 5 votes, B gets 4, and C gets 5. So, B is eliminated next.

# of voters:	5	4	3	2
1st choice:	A	B	C	C
2nd choice:	C	C	B	B
3rd choice:	B	A	A	A

We will skip rewriting the ballots and trust that you can see C will defeat A in the head-to-head runoff. A gets 5 votes, but literally all 9 other voters ranked A last, so A cannot get a majority.

All of this shows that C is 1st in the group ranking using IRV.

(b) Next, we will eliminate C from the ballots and apply IRV again. Think about how this is *different* from what we just did, where we were eliminating candidates one-by-one *while applying the IRV rules*. Here, we are eliminating C because we are moving to the second step in our quest to discover the group ranking.

Below are the ballots with C eliminated. We must apply the IRV rules to find the "winner" here. A gets 5 votes, B gets 4, and D gets 5. So, B will be eliminated first.

# of voters:	5	4	3	2
1st choice:	A	B	D	D
2nd choice:	B	D	B	B
3rd choice:	D	A	A	A

We trust that you can see D will defeat A in the head-to-head runoff. All of this shows that D is 2nd in the group ranking using IRV.

(c) Next, we eliminate D from the ballots. With just two candidates left, we can think of it as Majority Rules:

# of voters:	5	4	3	2
1st choice:	A	B	B	B
2nd choice:	B	A	A	A

Notice B wins 9-5, so B is 3rd in the group ranking.

(d) By default, this means A is 4th in the group ranking.

3. The group ranking using Top Two Runoff Voting will be $B > C > D > A$. (Notice that, again, the plurality winner A is *last* in this group ranking. Moreover, notice that B wins using this method, although B finished 3rd using the other two methods.)

(a) Let's apply the Top Two Runoff rules to the original ballots. Because A and B had the most 1st place votes, they advance to the runoff, while C and D are eliminated. We don't need to rewrite the ballots because you can see (right above this) the ballots reduced to just A and B. We see that B wins 9-5 over A in the runoff. This means B is 1st in the group ranking.

(b) Next, let's rewrite the ballots with B eliminated (table below) and apply the Top Two Runoff rules:

# of voters:	5	4	3	2
1st choice:	A	C	C	D
2nd choice:	C	D	D	C
3rd choice:	D	A	A	A

Candidates A and C have the most 1st place votes, while D is eliminated. In the runoff round, C defeats A 9-5. (Double check this by writing the ballots, if you need to convince yourself.) This means C is 2nd in the group ranking.

(c) Next, we eliminate C from all the ballots. This reduces to a Majority Rules contest between A and D. Candidate D wins that contest 9-5 (double check this), which means they are 3rd in the group ranking.

(d) By default, this means A is 4th in the group ranking.

Isn't that interesting! The three different voting methods yielded three different 1st place winners, and three completely different overall group rankings. This is further evidence of our point that *the voting method used matters as much as the ballots themselves.*
▲

2.5 Exercises

1. Below are tables of ranked ballots for three separate elections. To reinforce our knowledge of the voting methods we've learned so far, please do the following for each one:

   (a) Determine the winner using Plurality Voting.
   Then, determine the **group ranking** (Definition 2.4) using Plurality Voting. How, if at all, does this differ

from merely comparing each candidate's number of 1st place rankings?

(b) Similarly, determine the winner and then the group ranking using Instant Runoff Voting.

(c) Finally, determine the winner and then the group ranking using Top Two Runoff Voting.

Take note that Election 3 includes ballots that are *not full*, so you may have to consider how the total number of voters changes with each runoff round or each step in determining the group ranking.

Election 1:

# of voters	8	6	5	4	2
1st	B	A	C	D	C
2nd	C	C	A	A	D
3rd	D	B	D	B	A
4th	A	D	B	C	B

Election 2:

# of voters	8	4	4	3	2
1st	A	B	C	D	C
2nd	D	C	B	B	D
3rd	C	A	D	C	A
4th	B	D	A	A	B

Election 3:

# of voters	6	5	4	3	2	2	2	1	1	1
1st	C	A	B	C	A	D	B	A	D	B
2nd	B	D	A	A	D	B	A	B	-	-
3rd	-	C	D	D	B	-	-	-	-	-
4th	-	-	C	-	C	-	-	-	-	-

2. Consider the ballots from Election 1 in Exercise #1 above. Suppose two of the voters represented by the first column completely change their minds at the last minute and submit the ranking $A > B > C > D$ instead. Let's call that situation 1^* instead, and the table below shows the ballots:

Election 1^*:

	6	2	6	5	4	2
1st	B	A	A	C	D	C
2nd	C	B	C	A	A	D
3rd	D	C	B	D	B	A
4th	A	D	D	B	C	B

(a) Determine the winner of Election 1^* using the Plurality, Instant Runoff, and Top Two Runoff voting methods.

(b) How would you describe the ballot changes, going from Election 1 to Election 1^*? Are they *"in favor of"* A or not? Are they in favor of B? Or C? Or D?

(c) Reflect on your previous answers and describe anything you notice about the Plurality Voting results, comparing Elections 1 and 1^*. Does the change in outcome seem *"reasonable,"* considering the ballot changes and who they favored, or was it surprising in some way?

(d) Likewise, describe anything you notice about the Instant Runoff results. Does the change in outcome seem reasonable, or was it surprising in any way? How so?

(e) Finally, do the same for the Top Two Runoff results.

3. Shown below is a screenshot from *Ballotpedia* of the results for the 2020 general election for Maine's 2nd Congressional District in the U.S. House of Representatives [42]:

General election

General election for U.S. House Maine District 2				
The ranked-choice voting election was won by Jared Golden in round 1.				
Candidate	%	Total Votes	Transfer	Round eliminated
Jared Golden	53.0	197,974	0	Won (1)
Dale Crafts	47.0	175,228	0	1
Daniel Fowler	0.0	0	0	1
Timothy Hernandez	0.0	0	0	1
Undeclared Write-ins	0.0	33	0	

BP Incumbents are **bolded and underlined**. The results have been certified. Source   Total votes: 373,235

C = candidate completed the Ballotpedia Candidate Connection survey.

(a) Looking at the general election results, use the numbers in the *Total Votes* column to confirm the percentages in the first column.

(b) The general election called for Ranked Choice Voting to be used, but it appears there was only one round of voting. Why is that the case?

(By the way, the two names listed with 0 votes are most likely the names supplied by the 33 write-in votes.)

Now, here are the results of the Republican primary for that Congressional seat that was held earlier in the year [42]; notice there are two rounds:

Republican primary election

Republican Primary for U.S. House Maine District 2

Select round: Round 1 ✓

The following candidates advanced in the ranked-choice voting election: Dale Crafts in round 2 . The results of Round 1 are displayed below. To see the results of other rounds, use the dropdown menu above to select a round and the table will update.

	Candidate	%	Total Votes	Transfer	Round eliminated
	Dale Crafts	45.7	19,337	0	Advanced (2)
	Adrienne Bennett	31.8	13,468	0	2
	Eric Brakey	22.5	9,542	0	2

BP There were no incumbents in this race. Source   Total votes: 42,347
C = candidate completed the Ballotpedia Candidate Connection survey.

Republican primary election

Republican Primary for U.S. House Maine District 2

Select round: Round 2 ✓

The following candidates advanced in the ranked-choice voting election: Dale Crafts in round 2 . The results of Round 2 are displayed below. To see the results of other rounds, use the dropdown menu above to select a round and the table will update.

	Candidate	%	Total Votes	Transfer	Round eliminated
	Dale Crafts	58.5	22,888	3,551	Advanced (2)
	Adrienne Bennett	41.5	16,207	2,739	2
	Eric Brakey	0.0	0	-9,542	2

BP There were no incumbents in this race. Source   Total votes: 42,347
                                                                 Total exhausted votes: 3,252
C = candidate completed the Ballotpedia Candidate Connection survey.

(c) As before, use the *Total Votes* to confirm the percentages shown in the first column. Do this for both Round 1 and Round 2.

(d) In the results for Round 2, add together the three numbers in the *Transfer* column. Why is the result a negative number? What does this tell us about the voters who ranked Eric Brakey 1st in Round 1?

(e) In the results for Round 2, Dale Crafts has 58.5%. Is that his vote total as a percentage of all *original* ballots, or all *continuing* ballots? Calculate the other percentage and compare; before you do so, make a mental prediction of whether it will be smaller or larger than 58.5%.

(f) Is it possible for Adrienne Bennett to have won by assigning some of the *exhausted* ballots in Round 2 to be in her favor instead? If yes, what's the minimum number of those ballots that would need to be assigned to her? If not, explain why this is impossible.

(g) Forget the actual Round 2 results, and look back at Round 1. From there, is it possible for Adrienne Bennett to have won by assigning some of Eric Brakey's supporters to have Bennett ranked 2nd? If yes, what's the minimum number of those ballots that would need to be assigned to her? If not, explain why this is impossible.

4. Most of the examples in this chapter involved three or four candidates, so we want you to make up an example of an election with *five* candidates that produces completely different winners when we use Plurality, Instant Runoff, and Top Two Runoff. Is it even possible to make it so that there's a different winner when we use an analogous "Top Three Runoff" method?

5. With three candidates, there are 15 possible ranked ballots: 6 are full (ABC, ACB, BAC, BCA, CAB, CBA), 6 are not

full with two candidates ranked (*AB*, *AC*, *BA*, *BC*, *CA*, *CB*), and 3 are not full with one candidate ranked (*A*, *B*, *C*).

How many possible ranked ballots are there with four candidates? How many are full? How many have just three, two, or one candidates ranked?

And what about with five candidates? Can you generalize these results?

6. Find an op-ed in a newspaper (if possible, from your area) that argues for or against using Ranked Choice Voting. Then, look back at the Pros & Cons we discussed in Section 2.3 and try to identify whether any of the arguments in the editorial are similar to any of the ideas we shared.

   If you feel so inclined, consider writing a brief essay of your own that rebuts or agrees with the article you found.

7. Think of a situation in your own life where you might consider holding an election using runoffs. Write a proposal that describes the scenario and which voting method you think would work well. Use the ideas in this chapter, or anything you think of, to support your proposal. (And if this actually leads to a change, please contact us to tell us all about it!)

8. Identify a country that uses Instant Runoff or find an example of an election from recent history that used that method. (For inspiration, consult Wikipedia's *"History and use of instant-runoff voting."* [43])

   Then, create some kind of *"explainer"* that teaches someone about that country's voting method or about the results of that specific election, whatever it is you found. Think about a target audience and how you might best reach them. For example, you might make a short video to post on social media for your friends. Or, maybe a blog post that you can share with family. Or, a brief slides presentation for your student government or club executive board. Get creative!

The goal here is to teach the ideas of this chapter to someone else. If you can help *them* understand these topics, then *you* must really understand them well, yourself!

Chapter 3:
Methods of Pairwise Matchups, including Condorcet's Method

T H E goal of this chapter is to introduce you to a handful of related methods that all use **pairwise matchups** (or **pairwise comparisons**) to determine the outcome of an election. By this, we mean that a method uses the voters' rankings to compare candidates against each other one-on-one, and then makes a decision based on those matchups. The most famous historical example is **Condorcet's Method,** although it is more interesting from a mathematical perspective than a real world one. (It is not used in any country's elections.) Still, the ideas and methods in this chapter will be helpful in later investigations and merit some exploration and discussion.

3.1   Definitions and Examples

We continue to assume voters submit ranked ballots, and we will make two further assumptions for the rest of this chapter:

1. Voters submit **full ranked ballots**, meaning each voter ranks *all* the candidates in some order. This will be helpful so that

all one-on-one comparisons of candidates have the same *total number of votes*.

2. There are **no ties** in any matchups or results. In the real world, the *context of the election* would be used to break any ties, and it will be very convenient in all of our discussions here to just avoid those issues.

3.1.1 Pairwise Matchups

Now, before we define any new voting methods, let's use an example to clarify what we mean by **pairwise matchups**. Essentially, it means head-to-head comparisons (matchups) of two candidates at a time (pairwise).

Example 3.1. Let's revisit the ballots from Practice Problem 2.2, shown below.

# of voters:	5	4	3	2
1st choice:	A	B	C	D
2nd choice:	C	C	D	C
3rd choice:	B	D	B	B
4th choice:	D	A	A	A

1. *B defeats A in a pairwise matchup, by the count 9 to 5.*

   - *Translation:* If we ask the voters to ignore all the other candidates and focus only on how they rank A versus B, we would find that more voters prefer B.

   - *Another translation:* If we use the ranked ballots, ignore all candidates except A and B, and conduct a Majority Rules vote between just those two, B would win.

The five voters represented by the first column ranked A *somewhere above* B. For the sake of this pairwise matchup (A vs. B), it does not matter at all that A is 1st and B is 3rd; all that matters is A is *higher*.

Each of the other columns, though, has B ranked somewhere above A. (Indeed, they all have A ranked *last*, so B must be higher.) Those columns account for $4 + 3 + 2 = 9$ votes. Therefore, *B defeats A 9-5.*

2. *A loses all of their pairwise matchups.* What we just saw with A vs. B will occur with A vs. C and again with A vs. D. Candidate A gets 5 votes from the first column, but all 9 of the other voters ranked A last. So, C defeats A 9-5, and D defeats A 9-5.

   In this sense, A is a very poor candidate. If the electorate had been asked, "Do you like A or B more?", they would answer, "Collectively, we like B more!" The same thing would happen with A vs. C and A vs. D: the electorate prefers C over A, and prefers D over A. And yet, A is the Plurality winner simply because they have the most 1st place votes. So, this example shows it is quite possible to have a candidate that is hated by most of the electorate, but the other candidates split the 1st place votes in such a way that the hated candidate ekes out a slim plurality.

   To use some terminology we will introduce shortly, A is a **Condorcet loser** in this example.

3. *C wins all of their pairwise matchups.* At this point, we'll let you check our work to confirm these results:
   C defeats A 9-5, C defeats B 10-4, C defeats D 12-2.

In this sense, C is a strong candidate. Given the option of C or *someone else*, the electorate will always say, "We like C more," *no matter which candidate that "someone else" happens to be*. To use some terminology we will introduce shortly, C is a **Condorcet winner** in this example.

4. *B has a pairwise matchup record of 2-1, while D has a record of 1-2.* If you have not seen it before, this **Wins-Losses** notation is common, especially in sports. The phrase "B has a record of 2-1" means it has 2 wins and 1 loss. Indeed, in this example, B wins a pairwise matchup against A, loses against C, and wins against D.

Similarly, D only defeats A, and loses to B and C, hence the 1-2 record.

Using that idea, we could say that C is a Condorcet winner because they went 3-0, with no losses. And, we could say that A is a Condorcet loser because they went 0-3, with no wins.

3.1.2 Condorcet's Method: Win All Matchups

This method is named for a French philosopher and mathematician from the 1700s, the Marquis de Condorcet (pronounced roughly as *"con-door-SAY"*). After university, he worked on some mathematics in the new field of calculus, but later in life he focused on politics and philosophy. Specifically, he applied his mathematical skills to better understand political decisions. In fact, his discovery of the **Condorcet voting paradox**, which we will describe shortly, is often cited as the beginning of **social choice theory**, the formal scientific study of elections and decision-making.

Figure 1: Street sign in Paris for *Rue Condorcet*, honoring the Marquis de Condorcet. Source: photo taken by the author, March 2017.

Definition 3.1: Condorcet's Method.

Assume there are ≥ 3 candidates and voters submit ranked ballots (with no ties). The main idea is to use pairwise matchups and look for a candidate that wins all of their matchups.

1. *If there is a candidate that defeats each of the other candidates in a pairwise matchup, that candidate is declared the winner.*

2. *If no candidate defeats all the others in pairwise matchups (in other words, if every candidate loses at least one of their matchups), then there is declared to be no winner in the election.*

*The term **Condorcet winner** is used to refer to a candidate that happens to win all of their pairwise matchups in an election, even when Condorcet's method is not specifically being used to determine the winner of that election.*

That's it! It's a *demanding* method of pairwise matchups, asking for a candidate to win *all of their matchups*. So, it is quite common for there to actually be no overall winner, as we will see in some examples soon enough.

The comment about the term **Condorcet winner** means that we will sometimes refer to a "Condorcet winner" in an election to point out something interesting about that election, even if a different method (like Plurality or Instant Runoff) is being used to decide the actual winner. We may even have occasion to refer to a **Condorcet loser**, as well, just like we did above in Example 3.1. However, keep in mind that this is not quite the same as referring to a candidate that *did not happen to win under the Condorcet's Method rules*. (In that example, candidate C was the Condorcet winner because they won all their matchups. Candidates B and D won some and lost some of their matchups. But only candidate A was the *Condorcet loser* because they lost *all* their matchups.)

Example 3.2. Remember the example with nine friends choosing a cuisine? Let's apply Condorcet's Method to those ballots, shown on the right. With three candidates, there are three possible pairwise matchups to check.

# of voters:	4	3	2
1st choice:	I	P	T
2nd choice:	T	T	P
3rd choice:	P	I	I

- *I vs. T:* 4 voters rank I higher, the other $3+2=5$ voters rank T higher. So, T wins 5-4.

- *I vs. P:* 4 voters rank I higher, the other $3+2=5$ voters rank P higher. So, P wins 5-4.

- *P vs. T:* 3 voters rank P higher, the other $4+2=6$ voters rank T higher. So, T wins 6-3.

This means *Thai (T) is the Condorcet winner* because they won both of their pairwise matchups, defeating Indian (*I*) and Pizza (*P*). (It's also true that Indian is a Condorcet loser, going 0-2 in their matchups, and that Thai went 1-1 in their matchups.) So far, we have seen Indian win using Plurality Voting and Pizza win using Instant Runoff Voting. Now, we see that Thai wins using Condorcet's Method. Any of the candidates could win, depending on the voting method chosen!

How many possible pairwise matchups are there?

In that example, we said that three candidates meant there were three possible pairwise matchups. We don't want you to think those numbers are always equal! Here, we will briefly state a formula you can use to keep track of how many matchups there should be, based on the number of candidates. So, when applying Condorcet's Method to an example, you can make sure that you haven't missed any possibilities.

Theorem 3.1: Number of Pairwise Matchups.

Assume there are n candidates in an election, where $n \geq 3$, and the voters submit ranked ballots (with no ties). Then there are $M(n) = \frac{1}{2} \cdot n \cdot (n - 1)$ possible pairwise matchups of the candidates.

The tables on the right list the output of that formula for various values of n. (We use "$M(n)$" to stand for "*M*atchups with n candidates.")

n	$M(n)$	n	$M(n)$
3	3	8	28
4	6	9	36
5	10	10	45
6	15	11	55
7	21	12	66

Let's look at a specific value, say $n = 5$. Suppose the five candidates are A, B, C, D, E. We suggest listing possible pairs of candidates in "alphabetical order", like we do below. We see there are 10, confirming that $M(5) = 10$.

$$AB, \quad AC, \quad AD, \quad AE, \quad BC, \quad BD, \quad BE, \quad CD, \quad CE, \quad DE$$

Why does this formula work? Imagine you have n candidates and you are choosing a pair of them. There are n possible choices for the first candidate, and then there are $n - 1$ choices for the second candidate (one fewer, because you can't pick the first candidate twice). However, this *"double counts"* because it would account for selecting A then B, but it would also account for selecting B then A, and yet those are really the same matchup (A vs. B, the order does not matter). So, multiplying $n \cdot (n-1)$ counts the number of possible pairs, and then the $\frac{1}{2}$ in front fixes the double-counting.

Condorcet's Voting Paradox: a Non-Transitive Cycle

The word **transitive** is used in mathematics to mean that a relationship can *transition through* some middle ground. For example, when applied to numbers, "greater than" (symbolized by ">") is a transitive relation. If $x > y$ and $y > z$, it must also be true that $x > z$. The "greater than" relationship *transitions through* y in the middle.

The Marquis de Condorcet made a surprising discovery in the 1700s: **majority preferences can be *non*-transitive**. The simple example below shows how this is possible. There are three candidates (A, B, C) and three voters who rank those candidates in order. Let's determine the winner of each pairwise matchup.

80

# of voters:	1	1	1
1st choice:	A	B	C
2nd choice:	B	C	A
3rd choice:	C	A	B

- *A vs. B:* Two voters (left and right) rank A above B, so A wins 2-1.

- *A vs. C:* Two voters (middle and right) rank C above A, so C wins 2-1.

- *B vs. C:* Two voters (left and middle) rank B above C, so B wins 2-1.

There is no Condorcet winner here, nor is there a Condorcet loser. Every single candidate wins one of their matchups and loses the other one (meaning everyone has a 1-1 record). But there's another way to look at this: as a **cycle of majority preferences**.

Because A beats B in that one-on-one Majority Rules contest, we can say $A > B$. Likewise, $B > C$, and $C > A$. However, combining those three results, we get: $A > B > C > A$. Doesn't this violate the **transitive property** we just described? If $x > y$ and $y > z$, how can $z > x$ also be true? This example shows that majority preferences, based on ranked ballots, do not behave like standard numbers do!

This example also shows why Condorcet's Method is not commonly used in real world settings. The reason for holding an election in the first place is to *make a decision* and declare a winner. However, Condorcet's Method frequently yields an outcome where no candidate wins all of their matchups, meaning the official oucome is *"no definitive winner."* Not very desirable, in practice!

3.1.3 Copeland's Method: Compare Win/Loss Records

If the apparent flaw with Condorcet's Method is that it asks too much — a candidate must win *all* their matchups — then what about a method that asks a little less? **Copeland's Method** is a variation on Condorcet's Method that essentially looks for the candidate with the *best win/loss record* in their matchups. If a candidate wins all of theirs, then good for them: they will win under this method, too. But if there is *no Condorcet winner*, then this method may provide a "next best choice."

Definition 3.2: Copeland's Method.
Assume there are ≥ 3 candidates and voters submit ranked ballots (with no ties). The main idea is to use pairwise matchups and look for the candidate with the best win/loss record for their matchups.

1. *Determine the winner in every possible pairwise matchup.*

2. *For each candidate, count the number of matchups they win (call that number W), and the number of matchups they lose (call that number L). Their **Win/Loss Score** is those numbers subtracted: $W - L$.*

3. *The candidate with the highest Win/Loss Score is the overall winner.*

Example 3.3. Let's revisit the ballots from Example 3.1 (continuing from Practice Problem 2.2). We already computed the winner of each possible pairwise matchup, so here we can find each candidate's **Win/Loss Score**:

- A went 0-3, giving them a Win/Loss Score of $0 - 3 = -3$.

- B went 2-1, giving them a Win/Loss Score of $2 - 1 = 1$.

- C went 3-0, giving them a Win/Loss Score of $3 - 0 = 3$.

- D went 1-2, giving them a Win/Loss Score of $1 - 2 = -1$.

This confirms what we already saw in that election: C has the best record because they won all their matchups, B had the next best record, then D, then A. Also, notice that C had a *perfect score* because they were the Condorcet winner (won all their matchups), while A and D had *negative scores* because they lost more matchups than they won.

Historical Note: The Work of Raymond Llull

Interestingly enough, what we are calling "Copeland's Method" can be traced all the way back to the work of **Raymond Llull** in the 1200s(!) in Spain. He appears to be a fascinating person who worked in many different branches of knowledge, including mathematics, philosophy, and literature. His work on the theory of elections is among the earliest in human history, predating the Marquis de Condorcet by centuries. However, some of his manuscripts were only rediscovered in 2001, well after which many of the concepts and techniques contained therein had already been named for others. [44, 45]

Dealing with Ties in Copeland's Method

As we've mentioned before, we will usually not deal with *ties* in any way. However, you may be wondering right now: "What if there are ties in some *pairwise matchups*? Might that affect the *Win/Loss Score* somehow?" So, let's briefly discuss two issues regarding ties.

First, if there is a pairwise matchup that is tied, consider that as *neither a win nor a loss* for each candidate involved in that tie. In other words, it will neither increase nor decrease each candidate's

Win/Loss Score. This should make some sense: neither candidate defeated the other, so neither of them gets a boost in their score (and, likewise, neither of them gets penalized).

You might say, though: "Can't we give each candidate half a point for getting a tie, or something like that?" Sure, go for it! But you're not going to affect the *final outcome*; you're only going to possibly "bump up the numbers" involved. We will illustrate this idea in the example below so you see what we mean.

Second, what if there are candidates tied after computing their Win/Loss Score? Again, we will say that this boils down to a *context-dependent issue*, and whoever is holding the election should have a stipulation in place for how to handle ties in whatever method they've chosen. But there is a reasonable and interesting choice for a tie-breaking mechanism with Copeland's Method, so we will describe it here and then illustrate it with the example below. The idea is this: if two candidates tie for the highest Win/Loss Score, look at the candidates they defeated in their matchups, and add up the total Win/Loss Scores of *those* candidates. This is sort of like saying, "Well, we tied for the best record, but the opponents I beat had really good records, while the opponents you beat were not as strong, so I deserve to win."

Example 3.4 (Ties in Copeland's Method). Consider an election with ten voters and four candidates. The ballots are shown below, followed by the results of all six possible pairwise matchups and each candidate's Win/Loss (W/L) Record.

(*Study tip:* On your own, list all the matchups and calculate the winner of each one. Then, confirm your results with ours below.)

# of voters	3	3	2	2
1st	A	D	C	B
2nd	B	A	B	D
3rd	C	B	D	A
4th	D	C	A	C

1. A vs. B: A wins 6-4   • A's W/L Record: 2-1-0

2. A vs. C: A wins 8-2

   • B's W/L Record: 2-1-0

3. A vs. D: D wins 7-3

4. B vs. C: B wins 8-2

   • C's W/L Record: 0-2-1

5. B vs. D: B wins 7-3

6. C vs. D: Tied 5-5   • D's W/L Record: 1-1-1

In listing the Win/Loss Records above, we used the convention **Wins-Losses-Ties**. So, you can see that A and B are tied for the best record (2 wins, 1 loss), D is next (1 win, 1 loss, 1 tie), and C is worst (no wins, 2 losses, 1 tie).

According to Condorcet's Method, there is no winner here because no candidate won *all* their matchups. According to Copeland's Method, A and B are tied for the win. However, your intuition may be telling you that A deserves to win because, although A and B tied for the best record, A *defeated B in their matchup!* Indeed, this is precisely what the tie-breaking procedure we described above will confirm:

- A's Win/Loss Score: $2 - 0 = 2$

- B's Win/Loss Score: $2 - 0 = 2$

- C's Win/Loss Score: $0 - 2 = -2$

- D's Win/Loss Score: $1 - 1 = 0$

- A defeated B and C. The sum of their scores is $2 + (-2) = 0$

- B defeated C and D. The sum of their scores is $-2 + 0 = -2$.

- Because $0 > -2$, A wins the tie-breaker. Essentially, they defeated more challenging opponents overall.

3.1.4 Sequential Pairwise Method: "Knockout Agenda"

We could spend several chapters exploring variations on Condorcet's Method and the associated mathematics. Instead, let's look at just one more example because it will provide some motivation for a particular **property of voting methods** that we'll explore in the sequel to this book.

Definition 3.3: Sequential Pairwise Method.
*Assume there are ≥ 3 candidates and voters submit ranked ballots (with no ties). The main idea is to use pairwise matchups and follow a particular order of the candidates (called the **agenda**), facing them against each other one by one, like a "knockout tournament."*

1. *Before the ranked ballots are submitted, an **agenda** is established. This is a list of all the candidates in some order, with each candidate appearing exactly once in the list.*

2. *Take the first two candidates in the agenda and determine the winner of that matchup. The loser is eliminated and the winner advances.*

3. *Take that winner and match them against the next candidate on the agenda. Again, the loser of the matchup is eliminated and the winner advances.*

4. *Continue along the agenda, eliminating the loser and advancing the winner. The winner of the final matchup is the overall winner of the election.*

Hopefully, you can see the motivation behind the mouthful-of-a-name, **"Sequential Pairwise Method."** It indicates that pairwise (1-on-1) matchups are done by following a *sequence* (the agenda). But, if you'd prefer, you can think of this as the **"Knockout Method,"** as candidates are pitted against each other in the style of an elimination tournament (like the NCAA March Madness bracket).

It is also important to emphasize that *the agenda is chosen **before** the ranked ballots are submitted.* In other words, the rules of this voting method include having that list of candidates specified beforehand. Otherwise, if someone is able to choose the agenda *after* seeing the ballots, they might be able to tinker with the results and make their chosen candidate win. We will explore this idea in more detail in the example below.

Example 3.5 (The effect of the **agenda** in the Sequential Pairwise Method)**.** This example will show you how much the agenda matters in the Sequential Pairwise method. In fact, in this particular example, we will see that *each of the candidates could be the winner* by specifying an appropriate agenda.

The ranked ballots are shown below. (They're almost exactly the same as those in Example 3.4, but the first column has been changed to 4 voters instead of 3, so no matchups result in a tie.)

# of voters	4	3	2	2
1st	A	D	C	B
2nd	B	A	B	D
3rd	C	B	D	A
4th	D	C	A	C

Before moving on, confirm the results of all pairwise matchups:

- *A* vs. *B*: *A* wins 7-4
- *A* vs. *C*: *A* wins 7-4
- *A* vs. *D*: *D* wins 7-4

- *B* vs. *C*: *B* wins 9-2
- *B* vs. *D*: *B* wins 8-3
- *C* vs. *D*: *C* wins 6-5

Now, let's explore what happens with various agendas. If it would help, imagine that the candidates (*A*, *B*, *C*, *D*) are four possible pieces of legislation that could be enacted, and the eleven voters are members of a city council. In that situation, it is conceivable that the council meeting will have a specific agenda, in the more common sense of the word: a list of topics to discuss at the meeting, in a specified order. Imagine that the council will debate the pieces of legislation by considering two at a time and asking the council members to choose one or the other, then moving on to discuss that more popular piece of legislation alongside the next one on the list. (So, although it may seem strange to have some random order of the candidates as one of the "rules" of the voting method, it is perfectly natural in some real world settings like this.)

1. **Agenda: *A*, *B*, *C*, *D*. The winner will be candidate *D*.**

   - The first matchup is *A* vs. *B*, which *A* wins (7-4).

   - So, *A* advances to face *C*, next on the agenda, and *A* wins 7-4.

   - Finally, *A* faces *D*, and *D* wins 7-4.

2. **Agenda: *D*, *C*, *B*, *A*. The winner will be candidate *A*.**
   (Notice that this agenda is literally the reverse of the previous one, but it leads to a different outcome.)

   - The first matchup is *D* vs. *C*, which *D* wins 6-5.

- So, D advances to face B, next on the agenda, and B wins 9-2.

- Finally, B faces A, and A wins 6-5.

3. **Agenda: A, B, D, C. The winner will be candidate C.**

  - The first matchup is A vs. B, which A wins (7-4).

  - So, A advances to face D, next on the agenda, and D wins 7-4.

  - Finally, D faces C, and C wins 6-5.

4. **Agenda: A, D, B, C. The winner will be candidate B.**

  - The first matchup is A vs. D, which D wins 7-4.

  - So, D advances to face B, next on the agenda, and B wins 8-3.

  - Finally, B faces C, and B wins 9-2.

Notice that every single candidate *could* win: we showed one example of an agenda that leads to A winning, one that leads to B winning, and so on. This won't happen in *every election*, but it is worth showing that it *can happen*. Hopefully, this illustrates the fact that **the agenda matters a lot!**

If **there is a Condorcet Winner, then the Sequential Pairwise Method and Copeland's Method will** *always* **declare that candidate as the winner, too.**

It's worth pointing out that the two methods we just learned — Copeland's Method and the Sequential Pairwise Method — will *always identify the Condorcet Winner, when one exists*. Remember that we said the main problem with Condorcet's Method is that it is *too demanding:* it is common for no candidates to win all of their pairwise matchups, in which case Condorcet's Method simply says, "Sorry, I can't decide who wins!" However, *when there is* a Condorcet Winner, that candidate is so strong that they are *guaranteed* to be the winner under the other two methods we just mentioned.

To illustrate that, return to Example 3.2, the nine friends choosing from Indian, Pizza, and Thai food. In that example, we saw Thai (T) was the Condorcet Winner because they won both of their matchups: T went 2-0, Pizza (P) went 1-1, and Indian (I) went 0-2.

What if we had used Copeland's Method instead? The results would be the same: T gets a score of $2 - 0 = 2$, P gets $1 - 1 = 0$, and I gets $0 - 2 = -2$.

What if we had used the Sequential Pairwise Method instead? Well, doesn't it depend on the agenda? In this case, *no!* Try it yourself: choose a random agenda (list I, P, T in some order), and then conduct the knockout-style matchups. No matter what agenda you started with, T will emerge as the winning candidate. Because they win all their possible matchups, they are guaranteed to emerge from the knockout tournament unscathed: wherever they first appear in the agenda, they will win that matchup and all the rest.

Finally, there's nothing special about *this particular example.* It only mattered that T was the Condorcet Winner, so they were guaranteed to advance through every matchup along the agenda.

3.2 Real World Usage

As we've mentioned already, there are no real world examples of political elections that use Condorcet's Method. This is for what we hope is an obvious reason: the goal in holding an election is to make a collective decision, and yet Condorcet's Method sometimes declares no definitive winner!

Still, some voting methods have been created that use pairwise matchups to make a decision. Oftentimes, such a method is designed to essentially do this:

1. Look at all possible pairwise matchups. If a candidate wins *all* of theirs (as in, there's a *Condorcet Winner*), then they are declared the winner.

2. Otherwise, if there is *not* a Condorcet Winner, do *something else* with the results of the matchups to choose a winner.

That *something else* in Step 2 could be fairly simple, like in Copelugh's Method, or very complicated, like in something called the **Schulze Method**. The details of that method go well beyond the scope of this particular book, but we're mentioning it here because it has been adopted by several organizations in their decision-making processes, like software organizations (such as Debian and Gentoo) and some political organizations (the Pirate Party). [46, 47, 48]

3.3 Discussion of Pros and Cons

Although we just said that these methods are not very common in the real world, it is still worth discussing some of their benefits and disadvantages. This is because you may find yourself in a situation where a group of people needs to choose from a handful of options. It may be convenient for everyone to rank the options in order and then use those rankings to make a decision. These Pros and Cons below are really about using *pairwise comparisons* at all, and not necessarily about strictly following the rules of, say, Condorcet's Method.

3.3.1 Pros

- **Pairwise comparisons generalize Majority Rules.** With just two candidates in an election, Majority Rules is the only fair and reasonable voting method to use. But, with three candidates or more, there are many different voting methods one could use, as we are learning! In some sense, methods based on pairwise comparisons are the "most like" Majority Rules; that's what we mean when we say they "naturally generalize" the method of Majority Rules.

  Determining the winner of a pairwise matchup is conducting a miniature Majority Rules election inside a larger election. You use the voters' overall rankings of *all* the candidates and zoom in on just the two candidates you're matching up. So, many of these methods "inherit" the nice properties that the Majority Rules method has. (More on this below.)

- **It's reasonable to allow voters to have ties in their rankings and factor those in to the process.** Throughout this part of the book, and especially in this chapter, we're assuming that a voter's ballot is a *strict ranking* of the candidates. We aren't allowing a voter to say, "A and B are both my top choice, and C is 3rd." However, it's absolutely possible to allow that! This is especially true if you're conducting a small, informal election among friends, so you don't have many ballots to count.

- **Methods of pairwise comparisons have some nice theoretical properties.** Here, we mean *theoretical* in the sense of *according to the mathematical theory* of the subject. (We do *not* mean the more colloquial sense of the term *theoretical*, which is more like *hypothetically, possibly, abstractly*.)

Social choice theorists and mathematicians have developed a long list of different **criteria** or **properties** that describe voting methods and whether certain "weird situations" can occur in an election.

One of those properties is the **majority criterion**, which says that whenever a candidate is ranked first by more than half of the voters, then they must be the overall winner. Believe it or not, some voting methods *do not have this property!* (For example, the Borda Count, which we will see in Chapter 4, can possibly allow a candidate to win the election despite being ranked first on *nobody's* ballot.) Because methods like Condorcet's Method are designed to generalize from Majority Rules, they all obey the majority criterion, among several other nice properties.

3.3.2   Cons

- **Voters may be unable to express the *intensity* of their pairwise comparisons.** Imagine that you're making a decision among some friends, so you poll everyone about the three options. If one voter ranks $A > B > C$, perhaps they really feel strongly that A is their top choice, and B and C are a distant 2nd and 3rd. In other words, maybe the ">" in $A > B$ is a more intense comparison than the ">" in $B > C$. (And hey, maybe they really want A and don't care between B and C, but they ranked them that way on a whim when they submitted the ballot.)

It's possible to have an election where one candidate is the Condorcet winner (based on *pairwise matchups*), but a different candidate would win if those *intensities of comparisons* were counted. We have an example below in Section 3.4 to demonstrate this possibility.

Admittedly, this goes beyond Condorcet's method and other methods of pairwise comparisons. Truthfully, this could be considered a critique of *any kind of voting method that uses ranked ballots*. We are discussing this briefly now to motivate and preview Chapter 5 about voting methods that use *scores* instead of *ranks*.

- **Voters acting strategically at the individual level may cause a broadly disliked candidate to win the election overall.** Look back at Example 3.2 and the nine friends. Let's modify the ballots slightly to demonstrate something. Imagine there are 101 people voting instead, and 50 of them (not 4) ranked $I > T > P$, 50 of them (not 3) ranked $P > T > I$, and just one voter ranked $T > P > I$. This is meant to cartoonishly represent the situation where there are three viable candidates, two of which have broad support (almost half the electorate ranks Indian first, and almost half ranks Pizza first) while the third has very little support (the one voter who ranks Thai first).

In that situation, Thai would be the Condorcet winner, defeating each of the other two candidates 51 to 50 in a pairwise matchup. However, both other candidates are very close to having majority support! In that sense, isn't the contest really between Indian and Pizza? How could Thai sneak in and win like that? Perhaps the electorate acted in a way that seemed like a good strategy, but actually resulted in the worst possible outcome.

In debates leading up to actually casting ballots, perhaps the supporters of Indian realized that Pizza is their main rival.

Accordingly, they all decided to rank Pizza last on their ballots, hoping to ruin their chance of winning. Of course, this means putting Thai second on their rankings, even though nobody really wanted that choice to win. Likewise, perhaps the supporters of Pizza ranked their main rival, Indian, last on their ballots. While *each individual voter may feel like they are "acting strategically,"* the net effect is that the overwhelmingly least favorite candidate ended up winning.

This may seem like a contrived situation, and indeed the numbers were chosen to be nice. But situations like this can occur easily in real life, and it's very difficult to predict how voters will behave when they start to factor in *how **other** voters may rank the candidates* when they submit their own ballot. We are presenting this example now to demonstrate the idea of exploring possibilities; to demonstrate careful, logical thinking and analysis; and to preview some ideas in this book's sequel.

3.4   Practice Problems with Solutions

We have two examples to investigate. The first is designed to bring up some discussions about how to use pairwise comparisons to make an informed decision in the real world. This book is focused on exposing you to various voting methods and then exploring their mathematical properties in some detail. However, we don't want to lose sight of the fact that these voting methods are meant to be implemented in the world, to be used to make decisions!

Practice Problem 3.1. Suppose a teacher wants to create an activity for a future class meeting, and they want to incorporate student preferences in their decision-making process. The teacher decides on three reasonable options, which we will identify as the candidates A, B, C. (We trust you can imagine some specifics for this hypothetical situation if you want to make it more tangible.)

1. How should the teacher gather input from the students? Assume that the teacher has already decided they want to use pairwise comparisons of the candidates to help them make the decision.

2. Now, regardless of what you said for the first question, suppose that the teacher has gathered this information:

   (a) 11 students prefer option A. Eight of them said they would prefer option B next, but the rest did not give more specific preferences.

   (b) 8 students prefer option B. Of those, half mentioned they'd like option C next, but the other half did not give more specific preferences.

   (c) 6 students prefer option C. Of those, only one mentioned anything more specific, saying that they would probably like option B next.

   Organize this information into a ranked ballot table.

3. Use that ranked ballot table to determine the winner of every possible pairwise comparison.

4. Is there a Condorcet winner? Is there a Condorcet loser?

5. Which candidate "should" win, do you think?

6. Revisit the table of ranked ballots and look at the voters who only ranked one candidate. Is it possible to assign more specific rankings to those ballots and make a different candidate be the Condorcet winner?

▼ **Solution:** As we said, the point of this example is to remind us that elections are used to make decisions. It is not always the case that one should choose a specific voting method before gathering ballots.

1. In this hypothetical situation, it might be really strange and confusing for the teacher to put out an online poll that says: "Please rank the three options in your order of preference. All your rankings will be gathered and Copeland's Method will be applied to decide the winner." This is especially true if the students had never heard of Copeland's Method before!

   Moreover, there are practical concerns about how the logistics of gathering the information from students may influence their decisions. If a voting method is specified, perhaps students may engage in the strategic thinking we alluded to at the end of Section 3.3.2 that can led to paradoxical outcomes. The teacher may want to consider allowing students to assign numerical values to how strongly they feel. (Again, this previews some ideas from Chapter 5.) They should probably at least allow students to add written comments that the teacher can factor in to their decision.

   But, let's assume the teacher has decided the best choice for their context is to have the students rank the options and then use pairwise comparisons to choose an overall winner. Should they make an online survey where the students use multiple-choice questions to identify their 1st, 2nd, and 3rd place choices? This may not allow students to express indifference, as in: "I really like A, but B and C are the same to me." Should the teacher write down observations from their informal conversations with students? This might allow them to better capture their true preferences, but it might not give enough concrete data to work with.

In short, there are lots of possible things to consider, and they all depend on the context. But we encourage you to, if necessary, force yourself to *ask these kinds of questions.* When you need to make a decision on behalf of a group, think carefully about how you will gather information and how you will use it. And if you are participating in such decision-making, think carefully about how your ballot is gathered and how it might be used.

2. The student preferences we listed correspond to two types of ballots: one where a single candidate is ranked 1st and the others are left off, or one where all three candidates are ranked in order.

   For instance, we were told: *11 students prefer option A. Eight of them said they would prefer option B next, but the rest did not give more specific preferences.* This corresponds to eight ballots with the ranking $A > B > C$ and three ballots with the ranking A (and nothing in 2nd or 3rd).

   We trust you can read all that information again carefully and confirm that the table below presents the same data in a more organized manner:

# of voters	8	3	4	4	1	5
1st	A	A	B	B	C	C
2nd	B	-	C	-	B	-
3rd	C	-	A	-	A	-

3. With $n = 3$ candidates, there are 3 possible pairwise matchups.

- A vs. B: A wins 11 to 9.

The first column shows 8 voters ranking candidate A right above candidate B. The second column shows 3 voters ranking candidate A first. Although those voters did not list a 2nd choice, it is certainly reasonable to say that those 3 voters *prefer candidate A over candidate B*. Whether they would have put B 2nd or C 2nd, surely A would be ranked higher than B. So, we will count those 3 votes in favor of A for the sake of this matchup, giving A $8 + 3 = 11$ votes.

The five voters represented by the last column *do not specify* whether they prefer A or B. So, for the purposes of this pairwise matchup, it's as if those five voters are not part of the election.

The nine voters $(4 + 4 + 1)$ represented by the other columns all prefer B over A. All eight voters who rank B first are saying that they like B more than A. It does not matter, for instance, that four of those voters have not given specific preferences about A vs. C; whether A were 2nd or 3rd, it would certainly be below B.

- A vs. C: A wins 11 to 10.

The number of voters who rank A above C (whether explicitly or implicitly) is $8 + 3 = 11$. The number of voters who rank C above A is $4 + 1 + 5 = 10$. The remaining four voters ranked B first but were not specific about A vs. C.

- B vs. C: B wins 16 to 6.

Only three voters were not specific about B vs. C. Of the rest, only six of them ranked C above B.

Overall, notice that A won both of their matchups by slim margins. Moreover, neither of those wins were technically a *majority of all voters*, because a handful of voters did not give specific enough information. Meanwhile, although B only won one matchup, it was by a wide margin, wide enough to be a majority of *all* voters (not just the ones who provided enough information for that specific matchup).

4. Candidate A is the Condorcet winner in this election, according to Definition 3.1. However, we just pointed out that both of A's matchup victories are not technically *majority* wins. So, we understand if you feel like qualifying that statement to say, "A is the Condorcet winner, *if we allow* a candidate to have just a *plurality* victory in a matchup (not a *majority*)."

   Meanwhile, C is the Condorcet loser: they lost both of their matchups. In a similar vein, it's worth noting that one loss was a resounding one (losing to B, 16 to 6) but the other was close (losing to A, 11 to 10) and merely a plurality for the winner, at that.

5. If the voting method had been specified to be Condorcet's Method, A would be the winner. If it were Copeland's Method, A would be the winner again. And, hey, A is clearly the Plurality Voting winner here, with the most 1st place support. But is that the *best choice* for the teacher?

   Consider that B and A are pretty close in their pairwise matchup, and B trounces C head-to-head (in part because of all those voters who ranked A first but still listed B second). Meanwhile, A's pairwise matchups were only slim pluralities.

Overall, this is arguably a tough decision. There are compelling reasons to choose A, and there are compelling reasons to choose B. The margin of A's victory under Condorcet's Method is so slim that you may overturn the plurality vote and say that B is the best choice for the class at large.

Or, perhaps this is all evidence that C should not even be in the running, and that a group discussion should be held about the relative merits of options A and B before holding a second vote between just those two. It all depends on the context at hand, for sure, but after this kind of careful analysis and mathematical thinking, the teacher is likely better prepared to make an *informed decision*.

6. This final question addresses what we just alluded to about the slim victory by A under Condorcet's Method. It is possible to modify the ballots – not by changing anyone's ranking, but by imposing more specific information on some ballots – and make B the Condorcet Winner instead. This could be used as further evidence for B as the "best choice" for the teacher.

   In the table of ranked ballots, focus on the columns with only one candidate ranked first. For example, five voters put C first and nothing else. What if we made some of those ballots say $C > B > A$? Or $C > A > B$? We're not changing their top choice, and perhaps some voters would even supply those rankings if they were pressed further about their preferences. How could those changes affect the overall outcome?

   You can play around with some specific examples. For instance, what if all five of those voters ranked $C > A > B$? In that case, A would defeat B by a margin of 16 to 9 in their pairwise matchup. This would leave A as the Condorcet winner and give stronger support to it being the best overall choice.

But what if only some of those voters did that, and the rest left A and B unranked? And what about the other columns of voters? It quickly becomes too challenging to keep in mind all of the possible changes. In mathematical problem-solving, this is an indication that *we should define some new variables* to represent unknown numbers.

To keep this analysis a little more manageable, we will make this assumption right now: *all voters* must rank all the candidates in some order, with no ties and nothing unspecified. (Perhaps the teacher could state this rule on the original survey of the students.) This will just limit the number of variables to deal with and make the analysis a little easier.

- Of the three voters who only ranked A first, let's say x of them will actually rank $A > C > B$. This means the other $3 - x$ of them will rank $A > B > C$ (and get included in the first column). This also means that $0 \leq x \leq 3$; that is, x can only be 0 or 1 or 2 or 3.

- Of the four voters who only ranked B first, let's say y of them will actually rank $B > A > C$. This means the other $4 - y$ of them will rank $B > C > A$ (and get included in that column). This also means that $0 \leq y \leq$ 4.

- Of the five voters who only ranked C first, let's say z of them will actually rank $C > A > B$. This means the other $5 - z$ of them will rank $C > B > A$ (and get included in that column). This also means that $0 \leq z \leq 5$.

Based on what x, y, z are now defined to mean, make sure that you understand why this would be the new table of ranked ballots:

# of voters	$11 - x$	x	$8 - y$	y	$6 - z$	z
1st	A	A	B	B	C	C
2nd	B	C	C	A	B	A
3rd	C	B	A	C	A	B

Hopefully, you see why we chose to define x, y, z the way we did. It's easy to see here that, for instance, 11 voters ranked A first: $11 - x + x = 11$ is the total, and we have just split them based on who their 2nd choice was.

Let's recalculate all three pairwise matchups, *in terms of those unknowns*:

- A vs. B: $(11 - x) + x + z = 11 + z$ voters prefer A over B, while $(8 - y) + y + (6 - z) = 14 - z$ voters prefer B over A.

  (Hey, this accounts for all 25 voters, in total: $(11 + z) + (14 - z) = 25$.)

- A vs. C: $(11 - x) + x + y = 11 + y$ voters prefer A over C, while $(8 - y) + (6 - z) + z = 14 - y$ voters prefer C over A.

- B vs. C: $(11 - x) + (8 - y) + y = 19 - x$ voters prefer B over C, while $x + (6 - z) + z = 6 + x$ voters prefer C over B.

Now, what would it take to make B the Condorcet winner instead?

- *B must defeat A:* This means we would need $14 - z > 11 + z$ to be true. Subtract 11 from both sides, and add z to both sides; you get: $3 > 2z$. In other words, z can only be 0 or 1.

- *B must defeat C:* This is guaranteed! We would need $19 - x > 6 + x$. Subtract 6 from both sides, and add x to both sides; you get: $13 > 2x$. In other words, x can only be 0 or 1 or . . . up to 6. But we already know (when we defined x) that $0 \leq x \leq 3$ must be true anyway!

  You may have realized this already because, in the original election, B defeated C by a margin of 16 to 6. That only leaves 3 of the 25 votes unaccounted for, and the best hope for C only makes that margin 16 to 9 instead. In other words, B is guaranteed to beat C.

  But, it's worth taking a moment to recognize that we are using important concepts from **algebra** (defining variables to represent unknown values and writing things down using those variables) to reach that same conclusion. And moreover, there's a benefit to using variables like this: they will help us identify *all the possible ways* to make B the Condorcet winner, not just *whether it is possible*.

So, to make B the Condorcet winner, the only requirement is $z = 0$ or $z = 1$. (Meanwhile, x can be 0, 1, 2, or 3, and it doesn't affect B's chances of being the Condorcet winner.) In other words, of the voters who ranked C first, *at most one of them can rank A second!* If any more of them rank A second, then B has no chance of defeating A one-on-one.

We recommend reading over this example again, maybe even writing out the analysis for yourself where some of the voters still leave candidates unranked. For example, let's say x_1 is the number of $A > C > B$ ranks, x_2 is the number of $A > B > C$ ranks, and therefore $3 - x_1 - x_2$ is the number of ballots with just A ranked first. This may seem like overkill for such a "small example" where you can just play around with the possibilities by hand. However, the process of doing so, of practicing this kind of thinking, will help you analyze more complicated examples in the future.

▲

This second example is what we promised in Section 3.3.2: a situation where voters may express the *intensity* of their rankings by submitting scores for each candidate. We'll remind you that this is designed in part to preview Chapter 5, where we will explore similar voting methods in more detail. (The idea behind this example is partly inspired by a page on `RangeVoting.org` [49].)

Practice Problem 3.2. Suppose a decision must be made by three voters with four candidates. For example, maybe the voters are the leaders of a student club and they need to choose one of four different events to organize this semester. Rather than just ranking the options in order, the fact that there are so few voters leads them to decide that they can assign **numerical scores** to each candidate on a 1 to 10 scale.

The table below shows all of those scores, with one column per voter and one row per candidate. For instance, Voter 1 gave candidate A a maximum of 10 points, they gave 3 points to B, and they gave the minimum 1 point each to C and D. Note that there is no restriction on the *total* number of points a voter can give.

	Voter 1	Voter 2	Voter 3
A	10	1	6
B	3	2	10
C	1	4	8
D	1	10	1

1. What are the total scores for each candidate? Which one seems like the most reasonable choice for the winner of the election?

2. Use the scores provided to create a table of ranked ballots for this election. Determine the winner of each pairwise matchup and discuss what each voting method learned in this chapter would determine for this example.

▼ **Solution:** We will see that candidate A gets the most points when we factor in all the scores, but candidate B is the Condorcet winner if we ignore the scores and focus on just rankings.

1. It seems like a reasonable method is to simply add the scores assigned to each candidate and compare those totals:

   - A gets $10 + 1 + 6 = 17$ points.
   - B gets $3 + 2 + 10 = 15$ points.
   - C gets $1 + 4 + 8 = 13$ points.
   - D gets $1 + 10 + 1 = 12$ points.

   By this method, candidate A would be the winner with the highest total. Candidate B would be a close second, with C and D further behind. This seems like a reasonable outcome, given the voter's numerical scores for each candidate: A has one big supporter, one moderate supporter, and one strong opponent; meanwhile, B only has one strong supporter, C has at best middling support, and D has two voters firmly against them.

2. This next question asks us to explore what would happen if we *suppress* some of the information we have. For example, Voter 3's scores can be interpreted as the ranking $B > C > A > D$. Yes, this ignores some (possibly important) information that the voter provided, like the fact that D is five points behind A (comparing their 4th and 3rd choices) while A is only two points behind C (comparing their 3rd and 2nd choices).

   The point of this analysis is to show that *stripping away that information can indeed cause the outcome to be different.* But, conversely, this will also show that *adding new information, even if it is consistent with the old information, can change the outcome, too!*

   That is, if we think of what we're about to do with ranked ballots as the "original election," then allowing voters to describe their rankings by including scores (the table above) could be seen as *adding* information that would modify the overall outcome. So, this related pair of examples will show that it's possible both ways: including information can change the outcome, and excluding information can change the outcome.

   Below are the original ballots reinterpreted as rankings. Voter 1 gave both C and D 1 point, so they are technically tied for 3rd on Voter 1's ballot in the first column.

	Voter 1	Voter 2	Voter 3
1st	A	D	B
2nd	B	C	C
3rd	C, D	B	A
4th	-	A	D

With four candidates, there are six possible matchups:

- A vs. B: B wins, 2-1.
- A vs. C: C wins, 2-1.
- A vs. D: A wins, 2-1.
- B vs. C: B wins, 2-1.
- B vs. D: B wins, 2-1.
- C vs. D: tied, 1-1-1.

Overall, this shows that B is the Condorcet winner, defeating each other candidate in a head-to-head matchup.

How can we reconcile this with the earlier result? In the A vs. B matchup, the result "B wins, 2-1" ignores *the strength of each victory*. With only the ranked ballots, we would say two voters rank B higher than A. But, with the *numerical scores*, we can see that one voter ranked A *much higher* than B, while the other two ranked B *only slightly higher* than A.

To wit, the *score margin* of A over B is $(10 - 3) + (1 - 2) + (6 - 10) = 7 + (-1) + (-4) = 2$. This represents the fact that A is ranked higher than B by 7 points, and lower by 1 point and 4 points, for a combined margin of $+2$ points for A. This illustrates the difference between *ranked comparisons* and *score comparisons*, and it shows that the choice to include or exclude information from the process can have an effect on the outcome.

▲

3.5 Exercises

1. Look back to the three tables of ranked ballots in Exercise #1 in Section 2.5 on page #66. For each of those three elections:

   (a) Determine the winner of all pairwise matchups.

   (b) Identify if there is a Condorcet winner or a Condorcet loser. (Remember that there *could* be one, or the other, or neither, or both!)

   (c) Identify the winner using Copeland's Method. (If necessary, use the method described in Example 3.4 to break any ties.)

   (d) Then, find the group ranking (Definition 2.4) using Copeland's Method.

   (e) Finally, use all the pairwise matchup results you found in (a) to determine whether each candidate *could* be a winner using the Sequential Pairwise Voting method (Definition 3.3).

   That is, find an **agenda** that leads to A winning, or else explain why this is not possible. Then, do the same for B, then C, then D.

2. Look back to Example 2.4 and use that table of ranked ballots to answer the same questions in Exercise #1 right above this.

3. This exercise is based on the actual results of the 2009 mayoral election in Burlingon, Vermont, which used Instant Runoff Voting. There were three main candidates – Bob Kiss (K), Kurt Wright (W), and Andy Montroll (M) – and the tables below show the 8,833 ranked ballots that we will use (after two minor candidates were already eliminated). [50, 51]

# of voters	2043	371	568	1332	767
1st	K	K	K	M	M
2nd	M	W	-	K	W
3rd	W	M	-	W	K

# of voters	455	495	1513	1289
1st	M	W	W	W
2nd	-	K	M	-
3rd	-	M	K	-

(Note: These 9 columns were split into two tables merely to fit on the page. They collectively represent this one election.)

(a) In a scholarly article about this election, the authors write: *"Each of the three candidates has a legitimate claim to be the election winner."* [51] Is this a reasonable statement? Apply Plurality, Instant Runoff, and Condorcet's Method and describe what you find.

(b) In that same article, they demonstrate an *"upward monotonicity paradox"* in this example where Kiss could *lose* instead under Instant Runoff after some ballots are changed in his favor:

> *"In particular, what if 300 voters who had voted Wright, Kiss, Montroll and 450 voters who had voted Wright, −, − had instead voted Kiss, −, −?"* [51]

Rewrite the table of ballots to reflect these proposed changes. Then, apply Plurality, Instant Runoff, and Condorcet's Method to those ballots to determine the winners. Describe what you find.

(c) We have now seen that Instant Runoff Voting is susceptible to this kind of paradox, where moving a candidate *up* several voters' ballots can actually push that candidate *down* in the overall results. However, this is *not possible* under Condorcet's Method or Copeland's Method. Can you see why? Come up with a convincing, logical explanation for why ballot changes like those described in (b) cannot cause such paradoxical changes in the results under those methods.

4. Create an example of election results where there is a Condorcet loser and that candidate is also the Plurality Voting winner, or else come up with a logical explanation for why this is not possible.

5. In Example 3.4, we described a method for determining the winner under Copeland's Method when there's a tie for the best Win/Loss record. Can you create an example of election results where even *that* tie-breaking process leads to a further tie?

   If you find such an example, what *else* could be done to break the tie fairly? Can you invent a reasonable, logically consistent way to handle situations like that?

   Or, if you cannot find such an example, can you come up with a logical explanation for why this is not possible?

6. In Definition 2.2, when describing the runoff process for Instant Runoff Voting, we wrote: *"If there are two or more candidates tied for the fewest votes, they are all eliminated in this step."*

Now that we have discussed *ties* in this chapter, let's try amending that definition. Can you invent a reasonable, logically consistent way to handle situations like that and determine only *one* candidate to eliminate at a time? Consider using the results of pairwise matchups, even though the original definition of IRV didn't mention that idea. Then, apply your method to Example 2.5 and see whether the outcome changes at all.

7. Let's investigate how likely it is that a small election could lead to there being *no Condorcet winner*. To keep things as simple as possible, let's say there are just three voters (V1, V2, V3) and three candidates (A, B, C), and that voter V1 submits the ranking $A > B > C$.

   (a) How many possible *full* rankings could V2 and V3 submit together?

   (b) How many of those possibilities lead to *no* Condorcet winner?

   (c) Express these results as a **probability**. If the ballots for V2 and V3 were essentially chosen *at random*, what is the likelihood that we get an election with no Condorcet winner?

If you feel so inclined, consider extending this exercise to more voters. What if we had five voters and three candidates? How likely is it for there to be no Condorcet winner? Can you generalize even further?

Chapter 4:
Ranked Score Methods, e.g. Borda Count

ALL of the voting methods we have learned thus far are based on the voters ranking the candidates. You may have noticed, though, that the methods don't really "use all of the rankings" simultaneously. With Plurality Voting, only the 1st place rankings matter. With Instant Runoff and Top Two Runoff Voting, only the 1st place rankings are used to decide who will advance to the next round. And with Condorcet's and Copeland's Methods, the rankings are only used to compare two candidates at a time, and the *difference* between rankings doesn't matter: if I rank candidate A 1st and candidate B 3rd, those methods only "see" that I ranked A above B, not that A is *two* places above B.

So, you may be wondering if there are other voting methods that use the full rankings of the candidates and factor in their differences on each ballot. Yes, there are! We will call them **Ranked Score Methods** and the goal of this chapter is to introduce you to some of them. Our attention will mostly focus on the **Borda Count Voting Method** (usually shortened to just the **Borda Count**) because it is the most well-known and popular, and for good reason.

4.1 Definitions and Examples

The main idea behind all of the methods in this chapter is that candidates receive **points** from each voter based on where they are positioned on the voter's ranking. Then, each candidate gets a total **score** from all the voters, and the one with the highest score is declared the winner.

4.1.1 The Borda Count

Let's start with a simple example to demonstrate the main idea of the **Borda Count**. After that, we'll formally define that method, and then broaden the idea to define what a **Ranked Score Method** is, in general.

Example 4.1 (Illustrating the Borda Count). As always, let's return to the Indian/Pizza/Thai election, whose ballots are reproduced below on the right. Perhaps the friends have already realized that Indian is the Plurality Voting winner and Thai is the Condorcet Winner, but neither of those methods really use all of the information provided in the ballots. So, one friend suggests that they instead weight everyone's rankings using points, with a last place vote worth nothing and points going up by one for each rank above that:

- 1st place rank = 2 points.

- 2nd place rank = 1 point.

- 3rd place rank = 0 points.

# of voters:	4	3	2
1st choice:	I	P	T
2nd choice:	T	T	P
3rd choice:	P	I	I

114

Seems like a reasonable idea, right? Let's calculate the total score for each candidate. Don't forget to factor in the *number of voters* for each ballot!

- Indian gets 8 points.

  - The 4 voters in the left column rank I 1st, which is worth 2 points each. So, they get $4 \cdot 2 = 8$ points from those voters.

  - The 3 voters in the middle column rank I 3rd, which is worth 0 points each. So, they get $3 \cdot 0 = 0$ points from those voters.

  - The 2 voters in the right column rank I 3rd, which is worth 0 points each. So, they get $2 \cdot 0 = 0$ points from those voters.

  - Therefore, the total score for I is: $8 + 0 + 0 = 8$.

We could summarize that calculation as follows:

$$(4 \cdot 2) + (3 \cdot 0) + (2 \cdot 0) = 8 + 0 + 0 = 8$$

Notice how each term is (# of voters $\cdot$ # of points), which represents the total points given to the candidate from a whole block of voters who rank that candidate in the same spot on their ballot. Then, we add all those products together to get the candidate's total score. (We are coloring these blue and red throughout the example to make it easier to see.)

- Pizza gets 8 points:

$$(4 \cdot 0) + (3 \cdot 2) + (2 \cdot 1) = 0 + 6 + 2 = 8$$

- Thai gets 11 points:

$$(4 \cdot 1) + (3 \cdot 1) + (2 \cdot 2) = 4 + 3 + 4 = 11$$

Overall, Thai is the winner with 11 points (more than Indian and Pizza; they tied with 8). This result seems to factor in what you may have already noticed: Thai is ranked 1st by only two voters, but all the others ranked Thai 2nd. So, Thai picks up many points from those 7 voters. Meanwhile, Indian and Pizza are ranked last by several voters.

In that example, three candidates meant the points awarded based on each rank went 0, 1, 2. For more candidates, the idea is the same: 0, 1, 2, ..., up to *one less than the number of candidates* (because we started with 0). This is why a 1st place rank is worth $n - 1$ points in the formal definition below.

Definition 4.1: Borda Count Method.

Assume there are ≥ 3 candidates and voters submit ranked ballots (with no ties). Let's use n to mean the number of candidates (so $n \geq 3$). The main idea is to give candidates points based on where they are ranked on each ballot, and the candidate with the highest total score wins.

1. *If a candidate is ranked last (n-th place) on a ballot, they get 0 points from that voter.*

   If a candidate is ranked second-to-last ($(n - 1)$-th place) on a ballot, they get 1 point from that voter. And so on ... One rank higher on the ballot equals one more point from that voter.

   To put it another way: a 1st place ranking is worth $n - 1$ points, a 2nd place ranking is worth $n - 2$ points, and so on, down to 0 points for a last place ranking.

2. *The total number of points that a candidate receives from all the voters is called their **Borda Score**. The candidate with the largest Borda Score is the overall winner of the election.*

116

We pointed out how you can think of the point values in two ways: counting up from 0 at the bottom, or counting down from $n - 1$ at the top. Depending on your example, or how you see it, either of those might be more helpful for you. In particular, if a voter submits a ranked ballot that is *not full* (as in, they don't rank all the candidates), then it may help to count their scores from the top down instead of the bottom up, because they won't technically have a *last place ranking* on their non-full ballot.

Example 4.2. Let's practice applying the Borda Count to another example, this time with five candidates. We'll keep the discussion short; make sure that you agree with the calculated results by checking them on your own.

Shown below are the ranked ballots of 14 voters in an election with 5 candidates. Determine the winner of this election using the Borda Count method.

# of voters:	3	3	2	2	1	3
1st choice:	A	A	C	D	E	E
2nd choice:	B	E	B	C	A	B
3rd choice:	C	C	D	E	C	D
4th choice:	D	D	E	B	D	A
5th choice:	E	B	A	A	B	C

(Source: This example is a modified version of an example from *Chaotic Elections: A Mathematician Looks at Voting*, by Donald G. Saari [52]. The ballots actually shown on page 36 in that book are inconsistent with the claims made in the written text about the overall results of that election. We adjusted the ballots slightly – swapping pairs of candidates on just three ballots – to produce the ballots shown here, while maintaining the claimed outcomes.)

Brendan W. Sullivan

Definition 4.1 tells us that a 5th choice is worth 0 points, 4th is worth 1 point, 3rd is worth 2 points, 2nd is worth 3 points, and 1st is worth 4 points. As with Example 4.1, we will color the terms blue (# of voters) and red (# of points) so you can better see what's happening.

- Candidate A gets 30 points:

$$(3 \cdot 4) + (3 \cdot 4) + (2 \cdot 0) + (2 \cdot 0) + (1 \cdot 3) + (3 \cdot 1)$$
$$= 12 + 12 + 0 + 0 + 3 + 3 = 30$$

- Candidate B gets 26 points:

$$(3 \cdot 3) + (3 \cdot 0) + (2 \cdot 3) + (2 \cdot 1) + (1 \cdot 0) + (3 \cdot 3)$$
$$= 9 + 0 + 6 + 2 + 0 + 9 = 26$$

- Candidate C gets 28 points:

$$(3 \cdot 2) + (3 \cdot 2) + (2 \cdot 4) + (2 \cdot 3) + (1 \cdot 2) + (3 \cdot 0)$$
$$= 6 + 6 + 8 + 6 + 2 + 0 = 28$$

- Candidate D gets 25 points:

$$(3 \cdot 1) + (3 \cdot 1) + (2 \cdot 2) + (2 \cdot 4) + (1 \cdot 1) + (3 \cdot 2)$$
$$= 3 + 3 + 4 + 8 + 1 + 6 = 25$$

- Candidate E gets 31 points:

$$(3 \cdot 0) + (3 \cdot 3) + (2 \cdot 1) + (2 \cdot 2) + (1 \cdot 4) + (3 \cdot 4)$$
$$= 0 + 9 + 2 + 4 + 4 + 12 = 31$$

Therefore, E ekes out a slim victory with a Borda Score of 31 points, just barely above A's 30 points. In fact, all of the candidates' Borda Scores are separated by no more than 6 points, indicating a very close election overall. (Indeed, notice how each candidate appears near the top of many ballots and near the bottom of many ballots.) We will return to this example later in this chapter to apply other Ranked Score Methods to it.

4.1.2 The Points List of a Ranked Score Method

We can summarize how the points were awarded in Example 4.1 with $[2, 1, 0]$. There are three numbers in the list because there were three candidates: the first number is how many points a 1st place rank is worth, the second number is for a 2nd place rank, and the third number is for a 3rd place rank.

Similarly, we could summarize the points awarded in Example 4.2 with $[4, 3, 2, 1, 0]$. In general, when the Borda Count method is applied to an election with n candidates, the scores can be summarized with $[n-1, n-2, \ldots, 2, 1, 0]$. We will call something like this a **points list**, for hopefully obvious reasons. What ties together all the voting methods we're learning in this chapter? They all operate just like the Borda Count, but with different points lists.

Definition 4.2: Ranked Score Methods.

Assume there are ≥ 3 candidates and voters submit ranked ballots (with no ties). Let's use n to mean the number of candidates (so $n \geq 3$). The main idea is to give candidates points based on where they are ranked on each ballot, and the candidate with the highest total score wins.

*A **Ranked Score Method** is associated with its **Points List**, which represents how many points each possible ballot rank is worth to-*

wards a candidate's total score. That list must satisfy a few conditions:

1. *The list must have n numbers in it because there are n candidates and, therefore, n possible ranking positions on each ballot.*

2. *The list must end with a 0, meaning a last place rank is worth nothing. (See Example 4.3 for a reason why: essentially, it keeps the totals small.)*

3. *The list must be **non-increasing**, meaning that the numbers cannot get bigger as we read from left to right. (This is because we don't want a candidate to lose points from being ranked higher on someone's ballot.)*

Use the points list to award points to each candidate, based on where they are ranked on every single ballot. (For example, a 1st place rank is worth the number that is 1st in the points list; a 2nd place rank is worth the number that is 2nd in the list, and so on.) The overall winner is the candidate who gets the highest total number of points.

We hope you see how this is meant to generalize what we did with the Borda Count. There wasn't anything particularly special about the $[2, 1, 0]$ points list. We could have used $[4, 1, 0]$ instead, making a 1st place rank worth that much more than a 2nd place rank. Or, we could have used $[1, 1, 0]$, essentially weighting 1st and 2nd place ranks equally, but making 3rd place worth nothing. But it would be strange to use $[1, 2, 0]$ because that would mean a 2nd place rank is worth more than a 1st place rank. This is what we meant by **non-increasing** in Definition 4.2: reading left to right, it's okay for numbers to stay the same (like $[1, 1, 0]$) but it's *not* okay for numbers to increase (like $[1, 2, 0]$).

Why should the points list end with a 0?

Example 4.3. Let's revisit the Indian/Pizza/Thai election (see Example 4.1) and use the points list $[3, 2, 1]$ instead of the $[2, 1, 0]$ associated with the Borda Count. What would change? As it turns out: *nothing, really!* Every candidate's total score will increase, but *by the same amount*. So, the winner will be the same, and their margin of victory will be the same. Let's see how it works out:

- 1st place rank $= 3$ points.

- 2nd place rank $= 2$ points.

- 3rd place rank $= 1$ point.

# of voters:	4	3	2
1st choice:	I	P	T
2nd choice:	T	T	P
3rd choice:	P	I	I

- Candidate I: $(4 \cdot 3) + (3 \cdot 1) + (2 \cdot 1) = 12 + 3 + 2 = 17$

- Candidate P: $(4 \cdot 1) + (3 \cdot 3) + (2 \cdot 2) = 4 + 9 + 4 = 17$

- Candidate T: $(4 \cdot 2) + (3 \cdot 2) + (2 \cdot 3) = 8 + 6 + 6 = 20$

Remember the Borda Count results: Indian got 8, Pizza got 8, and Thai got 11. These are the same results, *just bumped up by 9 points each.* Why that number, why 9? Because there are 9 voters!

The points lists $[3, 2, 1]$ and $[2, 1, 0]$ differ by 1 point in every single corresponding number, so this is like each voter giving each candidate 1 extra point for free. It does not change the *relative scores* of the candidates (how they *compare*); it only changes the *absolute scores* (the total numbers).

This example illustrates why we stipulated, in Definition 4.1, that a points list must end with a 0. There's nothing inherently "incorrect" with ignoring that condition; you would just be dealing with larger numbers than you really need to.

Can the choice of points list affect who wins?

Yes, absolutely! Just like we keep saying that the choice of voting method matters as much as the ballots themselves, this idea applies specifically to Ranked Score Methods: the choice of the points list matters a lot, too. Let's keep using that Indian/Pizza/Thai example and see how the results can change.

Example 4.4. Let's try the points list $[4, 1, 0]$:

- 1st place rank = 4 points.

- 2nd place rank = 1 point.

- 3rd place rank = 0 points.

# of voters:	4	3	2
1st choice:	I	P	T
2nd choice:	T	T	P
3rd choice:	P	I	I

- Candidate I: $(4 \cdot 4) + (3 \cdot 0) + (2 \cdot 0) = 16 + 0 + 0 = 16$

- Candidate P: $(4 \cdot 0) + (3 \cdot 4) + (2 \cdot 1) = 0 + 12 + 2 = 14$

- Candidate T: $(4 \cdot 1) + (3 \cdot 1) + (2 \cdot 4) = 4 + 3 + 8 = 15$

In this case, Indian wins instead. The points list $[4, 1, 0]$ favors 1st place ranks more heavily than the Borda Count does, which boosts Indian's total score because they have the most 1st place votes.

Now, try the points list $[3, 2, 0]$ instead. We'll skip reprinting the ballots and showing the calculations, so make sure that you agree with these totals: Indian would get 12 points, Pizza would get 13, and Thai would get 20. In this case, Thai is the winner, just like with the Borda Count. However, the margin of victory is a bit wider than before (7 point difference, compared to only 3). Moreover, this "breaks the tie for 2nd," with Pizza edging out Indian by 1 point.

All of that happens because $[3, 2, 0]$ favors 2nd place votes a little more than the Borda Count does. With the Borda Count's $[2, 1, 0]$, a

1st place vote is twice as good as a 2nd place vote. But with $[3, 2, 0]$, a 1st place vote is only *one-and-a-half times* as good as a 2nd place vote. So, this helps out candidates that have 2nd place votes, like Thai (making them win by *more*) and Pizza (making them edge out Indian).

Try other points lists and see what happens! For instance, try $[4, 3, 0]$ and $[4, 2, 0]$. You should notice that $[4, 2, 0]$ yields the same results as the Borda Count, just with all the total scores doubled. (This is because $[2, 1, 0] \times 2 = [4, 2, 0]$, doubling every number in the list.) You also may have noticed that Thai wins with $[2, 1, 0]$ while Indian wins with $[4, 1, 0]$, so you may wonder if there is a points list that makes Pizza win. Can you find one? Try it!

Before we move on, we should point out that this idea of "assigning points to candidates" sounds like another idea that is actually quite different. Some voting methods give each voter a specified number of points to *distribute as they choose* among the candidates. For example, let's say I'm allowed 10 points to distribute: I could give them all to my favorite candidate, or I could split them with 8 points to one candidate and 2 to another, or I could give 2 points each to my 5 favorites, or ... This gives each voter a lot of freedom! However, with a Ranked Score Method, the points are determined entirely by the points list (which is set before the election) and where a voter puts a candidate on their ballot. That is the main difference. (We will learn about those other voting methods in Chapter 5.)

4.1.3 Other Named Ranked Score Methods

Some voting methods have descriptive names that let you guess how they operate. For instance, the **Vote for Two Method** works just like Plurality Voting, except each voter selects two candidates, not just one, and the winner is the candidate who gets chosen the most. (Note: A voter cannot pick the same candidate twice.) Now that we know about Ranked Score Methods, we can see that the Voter for Two Method is really a Ranked Score Method with the associated points list $[1, 1, 0, \ldots, 0]$. That is, it's like each voter giving 1 point each to their 1st choice and their 2nd choice, and 0 points to all the others.

In fact, the Plurality Voting Method is really a Ranked Score Method, too! It just happens to be a simple one whose points list is simply $[1, 0, \ldots, 0]$. A 1st place rank is worth 1 point, and all other ranks are worth nothing.

This section includes a couple of examples of voting methods that can be described using words (like the Vote for Two Method) but can also be thought of as Ranked Score Methods. We mention them here for completeness, so that you are introduced to them, but also to emphasize the connection to Ranked Score Methods, so you can better see their similarities.

Definition 4.3: Vote for Two (Three, Four, ...) Method.
Assume there are n candidates, with $n \geq 3$, and voters submit ranked ballots (with no ties). The main idea is to allow voters to select their top two (three, four, ...) candidates, and the candidate with the most votes wins.

- *For the Vote for Two Method, voters submit ranked ballots and we look at everyone's top two ranks only. (In practice, you don't need to ask for full rankings; you can just say, "Vote for your top two.") Each voter gives 1 point to each of those two candidates at the top of their ballot. The candidate with the most points/votes wins.*

This is a Ranked Score Method with points list $[1, 1, 0, \ldots, 0]$. (The "..." are there because we don't know exactly what n is, and we're saying, "After the first two 1s, it's all 0s.")

- *For the Vote for Three Method, voters submit ranked ballots and we look at everyone's top three ranks only. The candidate with the most such votes wins. This is a Ranked Score Method with the points list $[1, 1, 1, 0, \ldots, 0]$ (three 1s and then all 0s).*

- *This can be continued: for any value $k = 2, 3, \ldots, n - 1$, the "Vote for k Method" asks voters to choose their top k favorites. The candidate with the most such votes wins. This is a Ranked Score Method with points list $[1, \ldots, 1, 0, \ldots, 0]$, where there are k 1s followed by $n - k$ 0s.*

You may encounter these methods while reading about voting in other texts, so we want to ensure you understand the connection to Ranked Score Methods. The same goes for the next method which, although it has its own name, is actually already covered by the definition above.

Definition 4.4: Anti-Plurality Method.
*Assume there are n candidates, with $n \geq 3$. The **Anti-Plurality Method** is another name for the "Vote for $n - 1$ Method", as stated above in Definition 4.3. The main idea is to allow voters to specify which one candidate they're voting against, giving all the other candidates one vote each. Then, the overall winner is the candidate with the most total votes. This is equivalent to the Ranked Score Method with points list $[1, 1, \ldots, 1, 0]$, with $n - 1$ 1s and then a 0.*

Example 4.5. Let's apply these ranked score methods to the ballots from Example 4.2. We'll notice something interesting happen...

# of voters:	3	3	2	2	1	3
1st choice:	A	A	C	D	E	E
2nd choice:	B	E	B	C	A	B
3rd choice:	C	C	D	E	C	D
4th choice:	D	D	E	B	D	A
5th choice:	E	B	A	A	B	C

- Plurality Voting (Vote for One), or the points list $[1, 0, 0, 0, 0]$: A gets 6, B gets 0, C gets 2, D gets 2, E gets 4. A **wins!**

- Vote for Two, or the points list $[1, 1, 0, 0, 0]$: A gets 7, B gets 8, C gets 4, D gets 2, E gets 7. B **wins!** (To find these numbers, we are just looking in the top two rows and seeing whether a candidate appears in either spot. For example, E appears 2nd in a column with 3 voters, 1st in a column with 1 voter, and 1st in a column with 3 voters, for a total of $3 + 1 + 3 = 7$ votes.)

- Vote for Three, or the points list $[1, 1, 1, 0, 0]$: A gets 7, B gets 8, C gets 11, D gets 7, E gets 9. C **wins!**

- Vote for Four (Anti-Plurality), or the points list $[1, 1, 1, 1, 0]$: A gets 10, B gets 10, C gets 11, D gets 14, E gets 11. D **wins!**

- There is no "Vote for Five" because that would give all five candidates the same total score. Instead, we could use the Borda Count, which factors in all of the rankings, using the points list $[4, 3, 2, 1, 0]$. But we already saw in Example 4.2 that E **wins** using that method.

126

Look at that! Each candidate (A, B, C, D, E) won the election using one of those methods, including the Borda Count. In case you still needed convincing: *the choice of voting method matters as much as the ballots themselves!*

Quick comment about calculations: adding up the votes of all the candidates can help you catch errors. With Plurality Voting, the total votes should be 14, the number of voters: indeed, $6 + 0 + 2 + 2 + 4 = 14$. With the Vote for Two method, each voter is giving out 2 points, so the total should be twice as big: indeed, $7 + 8 + 4 + 2 + 7 = 28 = 14 \cdot 2$. You can use this to check your work. If the total doesn't add up right, look over your steps again carefully.

4.2   Real World Usage

4.2.1   Sports Competitions

Interestingly enough, the idea behind ranked score methods is used in several sports contexts.

Example 4.6 (Track Meets). Consider a sport that is individual in nature but has teams of individuals competing in a common tournament. For example, suppose we have a track meet where several schools each bring a team of runners to compete in several different events: the 100 meter dash, the 100 meter hurdles, the 400 meter dash, and the 1600 meter run. Based on a runner's placing in each of those events, they earn points for their team. At the end of the track meet, the team with the most total points wins.

Depending on the governing body and the number of teams involved, the points awarded by place vary slightly, but the main idea is the same. One particular guideline we found states the following [53]:

- With two teams competing in the meet, a 1st place finish in an event is worth 3 points, 2nd place is worth 2 points, 3rd place is worth 1 point, and anything lower is worth 0 points.

  This is like using the points list $[3, 2, 1, 0, \ldots, 0]$.

- With three or four teams, a 1st place finish is worth 5 points, a 2nd place finish is worth 3 points, a 3rd place finish is worth 1 point, and anything lower is worth 0 points.

  This is like using the points list $[5, 3, 1, 0, \ldots, 0]$.

- With four or more teams, 1st place is worth 10 points, 2nd is worth 8, 3rd is worth 6, 4th is worth 4, 5th is worth 2, 6th is worth 1 point, and anything lower is worth 0 points.

  This is like using the points list $[10, 8, 6, 4, 2, 1, 0, \ldots, 0]$.

You may not have thought of it this way before, but these kinds of track meets are like elections! The events are the voters and the teams competing in the meet are the candidates.

Each event (400 meter dash, 100 meter hurdles, etc.) is equivalent to one voter's ballot. That ballot has the candidates (the teams represented by the runners) ranked in order of how they placed in that event. Each team receives points based on their ranking. Then, the points for all the events are added together, just like we tally the points from all the ballots to determine the winner of an election. In this case, the points tell us which team finished 1st overall, which finished 2nd, and so on.

We should be careful, though, because there's one significant difference in this context. In a track meet, it is possible for a team to have one of their runners place 1st and another of their runners place 3rd, for example, in the same event. In terms of voting, this is like one voter (that event) *ranking the same candidate (team) both*

1st and 3rd on their ballot! This is obviously not allowed in the usual election contexts that we are considering in this book, but it can happen in a track meet.

This difference can have some major consequences for the theoretical understanding of these methods and for practical results. If we are in the standard election context (with voters and candidates), then the points list $[5, 3, 1]$ and the points list $[3, 2, 1]$ look different, although they will not produce fundamentally different outcomes, only different numerical totals. To see why:

- $[3, 2, 1]$ is equivalent to the Borda Count points list of $[2, 1, 0]$, as we explained in Example 4.3. Using $[3, 2, 1]$ instead just makes every candidate's total score larger (by exactly the number of voters, in fact).

- Then, $[2, 1, 0]$ is equivalent to $[4, 2, 0]$, as we mentioned at the end of Example 4.4. The points list $[4, 2, 0]$ is twice as big, in every single number, as $[2, 1, 0]$, so every candidate's score is just doubled. This will widen the margin of victory, but it will not change who finishes 1st (or 2nd or ...).

- Then, finally, $[4, 2, 0]$ is equivalent to $[5, 3, 1]$ for the very same reason that $[3, 2, 1]$ and $[2, 1, 0]$ are equivalent: the score for each ranking is just increased by 1, causing every candidate's score to simply increase by the number of voters.

So, the conclusion here is that in any standard election, the points lists $[5, 3, 1]$ and $[3, 2, 1]$ will always produce the same *overall ranking* of the candidates. They will produce different total scores, but the same relative rankings of those total scores. And by "standard election", we mean "a candidate can only appear once on a voter's ranked ballot."

However, in the context of track meets, *where a team can place 1st and 3rd simultaneously,* for example, those points lists can produce different overall winners! The key insight is recognizing how much a 1st place finish in an event is worth, relative to 2nd and 3rd place finishes combined. With $[3, 2, 1]$ scoring, a 1st place finish is equal to a 2nd and 3rd combined; but, with $[5, 3, 1]$ scoring, a 1st place finish is worth slightly more than a 2nd and 3rd combined. See one of the references [54] for an example of hypothetical race results that lead to different overall winners, using these $[5, 3, 1]$ and $[3, 2, 1]$ scoring methods.

Example 4.7 (Sports Team Rankings). The Wikipedia entry for the Borda Count [55] lists several other places where this method is used. For instance, they mention voting for certain sports awards (like the Heisman Trophy and MLB Most Valuable Player), as well as the NCAA AP Poll which is used to rank college sports teams:

> Each voter provides [their] own ranking of the top 25 teams, and the individual rankings are then combined to produce the national ranking by giving a team 25 points for a first place vote, 24 for a second place vote, and so on down to 1 point for a twenty-fifth place vote. [56]

That sounds like the points list $[25, 24, \dots, 2, 1]$, doesn't it?

Example 4.8 (Mario Kart). One more fun sports example: **Mario Kart!** A racing cup is like an election, where the racers are the candidates and each event in the cup is like a voter's ballot. Where a racer finishes in that event determines how many points they get from that ballot, and then the racers get ranked overall by how many points they earn throughout all the events. This author is most familiar with the Super Nintendo game *Super Mario Kart*, which uses the points list $[9, 6, 3, 1, 0, 0, 0, 0]$: with 8 drivers in each race, the top 4 places earn points and the bottom 4 places earn none.

Driver's Points

If a player comes in fourth place or better, they receive Driver's Points and will advance to the next course. Place fifth or lower and a menu will appear after you reach the goal. Select RETRY to restart the same course.

1st Place	2nd Place	3rd Place	4th Place
9 Points	6 Points	3 Points	1 Point

4th Place Or Better

5th Place Or Lower

When you play the 2-Player mode, if either player places fourth or better, both will advance to the next course.

Figure 1: An excerpt from the manual for *Mario Kart 64* showing the points awarded based on where a driver finishes in a race. Source: Screenshot of *MarioKart Wiki* [57]

4.2.2 Real World Examples of Elections

2015-2016 AP Football Poll

We mentioned the AP Poll for NCAA sports teams earlier in Example 4.7. Searching the internet for relevant examples yielded a Wolfram Demonstrations Project created by a few students for an undergraduate research project. [58]

This interactive tool allows you to explore the ballots and results of the 2015-2016 AP Football Poll, based on actual data. More specifically, you can modify the points awarded based on a voter's ranking and see how it affects the outcome. That is, you can essentially modify the points list, and it will automatically calculate the results for you!

Political Elections

The Wikipedia entry for the Borda Count also lists several places where this method (or something like it) is used in a nation's political elections, or in some other political context. [59]

Several universities and colleges use the Borda Count for electing student government officials or making other governing decisions. For example, Wheaton College in Massachusetts has used the Borda Count to select faculty committees. This process was studied by one of their mathematics professors, as well, which you can read about in the journal *Public Choice*. [60]

Moreover, there are a handful of countries that use some kind of Ranked Score Method in their political processes. One quite interesting example is the Pacific island nation of Nauru, whose method uses the points list $\left[1, \frac{1}{2}, \frac{1}{3}, \frac{1}{4}, \ldots\right]$. That is, a 1st place rank is worth 1 full point, and any ranks below that receive fractions of a point that decrease in value. Specifically, a candidate ranked in k-th place on a ballot receives $\frac{1}{k}$ points from that voter.

(Important caveat: Technically speaking, this is *not* a Ranked Score Method, according to our Definition 4.2! Remember that our definition requires a *"points list"* to end with a 0, meaning a last place rank is worth no points. However, this method instead makes a last place rank worth $1/n$ points, where n is the number of candidates overall. Still, we think this is *close enough* to methods like the Borda Count that we will continue to refer to it as a Ranked Score Method with a corresponding Points List.)

One of the consequences of this method's distribution of points is that it becomes much more challenging for a candidate to win the overall election with few 1st place ranks but lots of 2nd, 3rd, and lower ranks. Essentially, a 1st place rank is worth so much more than the rest that a candidate with the most 1st place votes tends to win anyway. Some academic studies have investigated this, including analyses of real-world election data and strategic bloc voting. [61, 62]

We mention this here to point out real world situations where Ranked Score Methods are used, but also to make sure you understand that the choice of points list impacts how the method operates in the real world. To help you understand, we will apply this **"Nauru Score Method"** to a few examples later on in Practice Problem 4.2.

4.3   Discussion of Pros and Cons

4.3.1   Pros

- **Ranked Score Methods factor in voters' full rankings of candidates, unlike other methods we have learned so far.** This is the same comment we made at the beginning of this chapter. All of the methods we have learned so far (plurality voting, runoff methods, and pairwise comparisons) arguably do not "use" every voter's full ranking to make a decision. However, Ranked Score Methods, like the Borda Count, do use every single voter's ranked ballot to contribute to the overall decision.

- **Voter familiarity with similar methods in the real world.** We mentioned a few real world contexts where *"point systems"* are used, like track meets. Because these are so similar to Ranked Score Methods, it's understandable that the general public may have more familiarity with and intuition for these voting methods (as compared to, say, Instant Runoff Voting). As usual, we'll point out that this is not a mathematical issue, but rather something to consider if you're advocating for a particular voting method to be used in a certain real world context.

- **The Borda Count has many nice theoretical properties, from a mathematical perspective.** This is a little vague, but that's on purpose so that we don't get too technical in this particular book (which is designed for a non-technical audience). But, in essence, the Borda Count is rather "nice", in terms of the theoretical, mathematical universe of voting methods. The mathematician Donald G. Saari is a notable proponent of the Borda Count, and Chapter 5 of his book *Chaotic Elections: A Mathematician Looks at Voting* can be read as an argument on behalf of the Borda Count over all other ranked score methods and over pairwise comparison methods. [52]

4.3.2 Cons

- **Potential for strategic voting.** This is the main downside of the Borda Count, and it is rather obvious and well-known, unfortunately. We have an example coming up in Section 4.4 to demonstrate what we mean by **strategic voting**, but the essence is this: a voter can submit a ranked ballot that does *not actually represent how they feel about the candidates* to ensure that the outcome of the election is *"better" for that voter* (as in, the overall winner is higher up on the voter's *true ranking* than would have occurred if they voted sincerely). It's not too difficult to construct examples that are ripe for this kind of tactical behavior, and this has even been studied and observed in real world contexts.

This is a good time to mention the origin of the name **Borda Count**. The method is named for **Jean-Charles de Borda**, a French mathematician, scientist, and Naval officer from the mid- to late-1700s. Borda constructed instruments for navigating the seas, based on mathematical principles, and he is well recognized in his native France for his contributions. His name is inscribed on the Eiffel Tower, for example, and several French Navy ships bear his name, too. [63]

Jean-Charles de Borda invented the method that is now named the Borda Count to solve a particular problem. (We should mention that the idea had already been discovered way back in the 1200s by Raymond Llull. We mentioned him in Section 3.1.3 for also inventing what is now known as Copeland's Method.) The context was electing officers to the French Academy of Sciences. [64] Borda created this method to be a fair and reasonable election method, but it was quickly pointed out to him (by the Marquis de Condorcet, in fact!) that the method can be susceptible to strategic voting. Borda's response was: "My scheme is intended for only honest men." In other words, this voting method's glaring flaw has been evident since its creation. [65]

Figure 2: Photograph of a statue of Jean-Charles de Borda on the esplanade of the French Naval Academy in Brest. Source: *Photos en Images* blog [66]

- **Potential for strategic nominations.** Closely related to the
  above point, the Borda Count and other ranked score methods
  may even encourage tactical behavior by political parties and
  factions, in terms of nominating candidates for an election.
  This is especially true when voters are *required to rank all
  the candidates* on their ballot, or else risk having their ballot
  entirely discarded. (This is true in Nauru, for example.) If
  a certain political party nominates several candidates, then
  their supporters can rank them all highly, allowing them to
  give *as few points as possible to their opponents.*

  Think of it this way: with just three candidates, the points
  list for the Borda Count is $[2, 1, 0]$. In that case, a 2nd place
  vote is worth *half* as much as a 1st place vote. Let's say one
  political party nominates two more candidates, so with five of
  them, the points list becomes $[4, 3, 2, 1, 0]$. Now, a supporter
  of that party can rank the three candidates from that party
  1st, 2nd, and 3rd in some order, pushing the other candidates
  to 4th or lower. Now, a 4th place vote is worth *one fourth* as
  much as a 1st place vote, which is quite a bit less than the
  $1/2$ with only three candidates.

4.4   Practice Problems with Solutions

This section includes three different examples because we have several different ideas we'd like to explore. The first example involves
just the Borda Count, but we'll see how changes on the voters'
ballots can cause seemingly bizarre changes in the overall results.
The second example involves the Nauru method with fractional
points that we mentioned in Section 4.2.2. And the third example
uses some basic algebra to show something interesting about the
Indian/Pizza/Thai example we keep exploring.

Practice Problem 4.1. This example will introduce you to some fundamental ideas that we'll explore in the sequel to this book about the mathematical properties of voting methods. Moreover, this makes good on our promise in Section 4.3 to show an example of **strategic voting**.

Consider the ranked ballots shown on the right for an election with 7 voters and 4 candidates. Throughout this example, we will use the Borda Count method, with the points list [3, 2, 1, 0].

# of voters:	3	2	2
1st choice:	A	B	C
2nd choice:	B	C	D
3rd choice:	C	D	A
4th choice:	D	A	B

1. Determine the overall winner of this election, using the Borda Count.

2. Suppose that the two voters in the middle column swap their rankings to be $B > D > C > A$ instead. Determine the winner. What do you notice?

3. Go back to the original ballots. Suppose that the two voters in the right column swap their rankings to be $C > D > B > A$ instead. Determine the winner. What do you notice?

4. Go back to the original ballots. Suppose that candidate D drops out of the race at the last moment, so the ballots are "condensed" to have just three candidates and three ranks. Determine the winner, again using the Borda Count (with [2, 1, 0] now). What do you notice?

▼ **Solution:** As you may have guessed, the ballot changes introduced in the later questions will yield different outcomes of the election. Let's see ...

1. In the original election, A receives $9 + 0 + 2 = 11$ points, B receives $6 + 6 + 0 = 12$, C receives $3 + 4 + 6 = 13$, and D receives $0 + 2 + 4 = 6$. Thus, C is the overall winner.

2. The two voters in the middle column are swapping their rankings from $B > C > D > A$ to $B > D > C > A$. (The new ballots are shown below.) The only changes involve C and D (marked in red). It looks like C will lose some points and D will gain some (but, meanwhile, A's and B's Borda Scores will not change). In fact, C will lose exactly 2 points and D will gain those 2 points. This is because we're changing two ballots, and D moves up one rank while C moves down one rank, which corresponds to a change of *1 point per ballot*.

- A gets $9 + 0 + 2 = 11$
- B gets $6 + 6 + 0 = 12$
- C gets $3 + 2 + 6 = 11$
- D gets $0 + 4 + 4 = 8$

The winner is now B!

# of voters:	3	2	2
1st choice:	A	B	C
2nd choice:	B	D	D
3rd choice:	C	C	A
4th choice:	D	A	B

Here's how this example relates to **strategic voting**. Imagine that you are one of those two voters in the middle column. Your top choice, candidate B, was very close to winning the original election: they were only one point below the winner, candidate C. Now, C was your 2nd choice, so you might not be upset with the result, but you certainly would have preferred for B to win instead. You can help make that happen by *pushing C down your ballot!* That is, you can *submit* *an **insincere ballot*** that does not reflect your "true personal preferences" about the candidates.

Indeed, in this particular example, if *both* of those voters do that, they take two points away from candidate C. Because C originally won over B by just one point, they will now *lose by one point*, and that's exactly what we see in the results above. This is what **strategic voting** means: those two voters submit ballots that don't reflect their true preferences, and yet this leads to a *more preferable outcome for those voters* (as in, the new winner is higher on their true ranking).

Now, this seems to require the voters to have some intimate knowledge of everyone else's ballots, and it also seems to require some organized behavior by the voters. (If only *one* of the voters modified their ballot, it would make C and B tie instead.) So, it's debatable how worrisome this phenomenon is for real world election contexts. However, this example here at least demonstrates that such a situation is *possible*.

3. Let's rewrite the ballots, with the changes marked in red, and calculate the new Borda Scores:

- A gets $9 + 0 + 0 = 9$
- B gets $6 + 6 + 2 = 14$
- C gets $3 + 4 + 6 = 13$
- D gets $0 + 2 + 4 = 6$

The winner is now B!

# of voters:	3	2	2
1st choice:	A	B	C
2nd choice:	B	C	D
3rd choice:	C	D	B
4th choice:	D	A	A

Just like with the previous question, we have some ballot changes leading to B winning instead of the original victor

C. This time, though, it's not because of strategic behavior by the voters. Instead, one can argue that this situation is strange and undesirable.

Originally, C was the winner. In the new set of ballots, the only changes affect *the pairwise comparisons of A versus B, but no other pairs*. That is, if we look at the original ballots, everyone who ranked C above B *still* ranks C above B (and vice-versa) in the new ballots. Likewise, everyone who ranked B above D in the original ballots *still* ranks B above D in the new ballots. The same goes for all possible pairs, except A versus B.

In the new ballots, two voters are changing their mind about how they compare candidates A and B: specifically, they are now saying, "We like B more than A, instead of A more than B." Isn't it a little strange that *this will cause C to lose?* We can see how the numbers work out: in the new ballots, B gains 2 points (1 point from each of the 2 voters), and this leapfrogs their score ahead of C's score. But it doesn't quite make sense in terms of the preferences of the group. Originally, C beat B. Then, nobody changed their mind about how they rank C versus B. And yet, B jumped ahead of C in the overall results.

This one example demonstrates that the Borda Count voting method *fails the **Independence of Irrelevant Alterantives** property*. We will learn about this property in a sequel to this book, but we wanted to preview the idea now to make it easier to understand later on, and to encourage you to think about this phenomenon in the meantime.

4. Finally, let's see what kind of weirdness can occur when a candidate is removed from the ballots. In this case, let's

140

rewrite the ballots without candidate D and calculate the $[2, 1, 0]$ Borda Scores:

# of voters:	3	2	2
1st choice:	A	B	C
2nd choice:	B	C	A
3rd choice:	C	A	B

- A gets $6 + 0 + 2 = 8$
- B gets $3 + 4 + 0 = 7$
- C gets $0 + 2 + 4 = 6$

The winner is now A!

This one may be the most surprising of the three situations in this example because there are no ballot *changes*, per se. No voter is changing their mind or anything like that. We are merely taking their already stated preferences and ignoring one of the candidates. Still, this caused the results to be different. (Not only does A win instead of C, this time, but C even drops to 3rd place, below B!)

As we will discuss later, this is actually another way to interpret that **Independence of Irrelevant Alternatives** property. Imagine that all of these seven voters have their preferences about the candidates, as stated in their original ballots. Apparently, taking out candidate D, or including them, can change the **group ranking** of the other candidates, A and B and C. This indicates that the voting method in question (the Borda Count) is somehow sensitive to these kinds of changes. That is, the results are *dependent* on whether or not an *alternative* (candidate D) is included.

This may have significant real world implications! For example, imagine that this election was held for the position of Student Government President. Four candidates ran, and C was going to be declared the winner. But after ballots were cast yet before the results were announced, candidate D said, "I'm dropping out of the race." (Perhaps they accepted a part-time job that would interfere with their duties, or something like that.) Should the results be based on the voter's original ballots, making C the winner? Or should candidate D be removed from the ballots first, leading to A as the winner? Regardless of the actual decision made, it certainly seems important for the organizers of the election to at least be aware that such a decision needs to be made and that, furthermore, it could even influence the results.

(Examples like this — where small changes on the ballot cause significant changes in the overall outcome — will be common when we investigate the mathematical properties of voting methods.)

▲

Practice Problem 4.2. In Section 4.2.2, we mentioned that Nauru uses a Ranked Score Method with the points list $\left[1, \frac{1}{2}, \frac{1}{3}, \frac{1}{4}, \ldots\right]$. For brevity, let's call this the **Nauru Score Method**. Apply this method to the various examples we've seen in this chapter, whose ballots are reproduced below for convenience.

Election 1:

# of voters:	4	3	2
1st choice:	I	P	T
2nd choice:	T	T	P
3rd choice:	P	I	I

Election 2:

# of voters:	3	2	2
1st choice:	A	B	C
2nd choice:	B	C	D
3rd choice:	C	D	A
4th choice:	D	A	B

Election 3:

# of voters:	3	3	2	2	1	3
1st choice:	A	A	C	D	E	E
2nd choice:	B	E	B	C	A	B
3rd choice:	C	C	D	E	C	D
4th choice:	D	D	E	B	D	A
5th choice:	E	B	A	A	B	C

▼ **Solution:** We will show the calculations and state the results for these elections. We encourage you to attempt this *on your own first*, and then consult our explanations to check your work.

1. There are three candidates, so we use the points list $\left[1, \frac{1}{2}, \frac{1}{3}\right]$.

   (a) Candidate I receives roughly 5.67 points:

   $$4{\cdot}1+3{\cdot}\frac{1}{3}+2{\cdot}\frac{1}{3} = 4+1+\frac{2}{3} = 5+\frac{2}{3} = \frac{17}{3} = 5.6666\ldots$$

   (b) Candidate P receives roughly 5.33 points:

   $$4\cdot\frac{1}{3} + 3\cdot1 + 2\cdot\frac{1}{2} = \frac{4}{3}+3+1 = 5+\frac{1}{3} = 5.3333\ldots$$

   (c) Candidate T receives 5.50 points:

   $$4\cdot\frac{1}{2} + 3\cdot\frac{1}{2} + 2\cdot1 = 2+\frac{3}{2}+2 = 5+\frac{1}{2} = 5.5$$

Overall, Indian is the winner, although the scores are all pretty close! As we stated in Section 4.2.2 when introducing this Nauru method, it tends to elect candidates with many first place votes, even if they don't have broad support. Indeed, in this example, we know that Thai wins the Borda

Count because of its seven 2nd place ranks. With this Nauru method, though, that's still not enough to overcome Indian's four 1st place ranks.

2. With four candidates, we use the points list $\left[1, \frac{1}{2}, \frac{1}{3}, \frac{1}{4}\right]$.

   (a) Candidate A receives roughly 4.17 points:

   $$3 \cdot 1 + 2 \cdot \frac{1}{4} + 2 \cdot \frac{1}{3} = 3 + \frac{1}{2} + \frac{2}{3} = 4 + \frac{1}{6} = 4.1666\ldots$$

   (b) Candidate B receives 4.00 points:

   $$3 \cdot \frac{1}{2} + 2 \cdot 1 + 2 \cdot \frac{1}{4} = \frac{3}{2} + 2 + \frac{1}{2} = 4$$

   (c) Candidate C receives 4.00 points:

   $$3 \cdot \frac{1}{3} + 2 \cdot \frac{1}{2} + 2 \cdot 1 = 1 + 1 + 2 = 4$$

   (d) Candidate D receives roughly 2.42 points:

   $$3 \cdot \frac{1}{4} + 2 \cdot \frac{1}{3} + 2 \cdot \frac{1}{2} = \frac{3}{4} + \frac{2}{3} + 1 = 2 + \frac{5}{12} = 2.4166\ldots$$

   This time, A squeaks out a slim victory over B and C, who tie for 2nd. With the Borda Count, we saw that C won, thanks to being ranked only 1st, 2nd, and 3rd. This time, we see the Plurality Voting winner emerging as the victor, just like the previous example, even though they are also ranked last by a couple of voters.

3. Now that we've seen some examples, we will skip the results for this third election, and leave it to you to verify the calculations. (It's good practice!) Remember to use the points list $\left[1, \frac{1}{2}, \frac{1}{3}, \frac{1}{4}, \frac{1}{5}\right]$. (Whenever we say *"roughly"*, that means we rounded a total to two decimal places.)

(a) Candidate A receives 8.05 points.

(b) Candidate B receives 5.30 points.

(c) Candidate C receives roughly 5.93 points.

(d) Candidate D receives roughly 5.42 points.

(e) Candidate E receives roughly 7.27 points.

Overall, this makes A the winner, although we saw that E won using the Borda Count. This time, A's six 1st place votes help them earn more points than E, who only had four 1st place votes.

▲

Practice Problem 4.3. Let's revisit the Indian/Pizza/Thai election from Example 4.4, where we showed that the choice of points list can affect the overall outcome. Specifically, we had already seen (Example 4.1) that Thai wins using the Borda Count with points list $[2, 1, 0]$. Then, we saw that Indian would win using the points list $[4, 1, 0]$, which more heavily weights 1st place ranks.

At the end of that example, we encouraged you to try and find a points list that makes Pizza the winner. Here, we will use some simple algebra to show that, in fact, *it is impossible for Pizza to win* using *any* points list!

▼ **Solution:** Remind yourself of the stipulations for a points list, given in Definition 4.2. With $n = 3$ candidates in this example (ballots shown below), we must have a list of three numbers that ends in a zero and that is *non-increasing*. That's all we know. Our goal here is to explore how the winner of the election is affected by the choice of the first two numbers in the points list. We could do this by *"guessing and checking,"* essentially playing around with a bunch of examples. That's perfectly fine, and we encouraged you to do so! However, we would now like to be *methodical* and efficient with those explorations. And because we have some unknown values to think about (the numbers in the points list), this is precisely the point where a mathematical mind says, "Aha, let's *define some variables!*"

So, we will define the generic points list $[a, b, 0]$ for this example. We chose "a" and "b" just because they're the first two letters. (You could use x and y, or v_1 and v_2, or even $\heartsuit$ and $\diamondsuit$, if you'd like.) The *non-increasing stipulation* means that $a \geq b$ must be true: we could have $a = b$ (meaning 1st and 2nd are worth equal points), or we could have $a > b$.

We've already seen that $a = 2$ and $b = 1$ (the normal Borda Count) makes Thai win, and $a = 4$ and $b = 1$ makes Indian win. Now, the main question is this: *Are there any values of a and b that make Pizza win instead?*

Look at the ranked ballots of the nine voters. Using the points list $[a, b, 0]$, we can write down the total score for each candidate **in terms of** those unknown values a and b. (Remember that phrase, "in terms of," from math class? It just means, "using those variables to represent unknowns, not necessarily writing a specific number.") Make sure you can confirm these results:

# of voters:	4	3	2
1st choice:	I	P	T
2nd choice:	T	T	P
3rd choice:	P	I	I

- Indian gets $4a$ points.

- Pizza gets $3a + 2b$ points.

- Thai gets $2a + 7b$ points.

So, what will it take for Pizza to be the winner? Their score must be larger than the other two scores. In other words (and symbols):

$3a + 2b > 4a$ must be true, and $3a + 2b > 2a + 7b$ must be true.

Let's take those **inequalities** one by one. The first one says $3a + 2b > 4a$. If we subtract $3a$ from both sides, we get $2b > a$. In other words, to make sure Pizza (with $3a+2b$ points) defeats Indian (with $4a$) points, a has to be less than $2b$. This means that a 1st place score can't be twice as good as a 2nd place score. That should make some sense, looking at the ballots: Indian has many 1st place votes, so to give Pizza a chance, we would need 2nd place votes (of which Indian has *none!*) to be worth enough.

Based on the requirement $a < 2b$, perhaps $[3, 2, 0]$ would work as a points list. With $a = 3$ and $b = 2$, we have $a < 2b$ because $3 < 4$. Does that make Pizza win? Compute the scores for yourself: you will find that Pizza does indeed beat Indian, but Pizza loses to Thai. It seems like we need to consider that other inequality which must *simultaneously* be true to make Pizza the overall winner.

Let's take $3a + 2b > 2a + 7b$ and subtract $2a$ from both sides. That produces $a + 2b > 7b$. Next, let's subtract $2b$ from both sides to get $a > 5b$. This says that a 1st place score must be worth much more than a 2nd place score. (In fact, it says 1st must be worth *at least five times as much* as 2nd.) That makes sense when we look at the ballots: Thai has *seven 2nd place votes*, while Pizza has only two. So, to give Pizza a chance at defeating Thai, we would need 1st place votes to be worth *a lot* (because Pizza has one more than Thai does).

Unfortunately for Pizza's supporters, it is not possible to satisfy *both of those requirements at the same time.* One says $a < 2b$ and the other says $a > 5b$. But how can those both be true? How can a be smaller than $2b$ but also bigger than $5b$, which is bigger than $2b$ itself?! (More strikingly: $5b < a < 2b$ implies $5b < 2b$, which is impossible when $b \geq 0$.)

If you played around with this example on your own, testing out various points lists, you may have noticed this effect already. You may have realized that 1st place votes can't be worth too much (or else that makes Indian win), but 1st place votes also have to be worth a lot (or else Thai would win). Maybe you felt that but couldn't quite put it into words or symbols. But now that we're here, read over this example again. Think about how we used variables — symbols to represent unknown values — to express those some observations using concrete, mathematical language and symbols. *This is the power of mathematical thinking.* ▲

4.5 Exercises

1. Look back to the tables of ranked ballots in Exercise #1 in Section 2.5 on page #66. For each of Election 1 and Election 2:

   (a) Determine the winner using the Borda Count.

   (b) Determine the group ranking (Definition 2.4) using the Borda Count.

   (c) Make up a points list other than [3,2,1,0] and apply that ranked score method. Was there a different winner? A different group ranking?

   (d) If possible, determine whether each candidate could possibly win by using a different points list. Could A be the overall winner? How, or why not? And what about B? Or C? Or D?

   (e) For extra practice, apply the **Nauru Score Method** (see Problem 4.2), as well, to determine the overall winner and the group ranking.

   Finally, speculate on why we didn't ask you to do anything with Election 3. What is different about that election that would make applying the Borda Count or other ranked score methods somehow "unfair" or misguided? How might the voters have acted differently if they knew that putting a candidate *somewhere* on their ballot would give them points? And based on all this, can you see why Nauru requires a voter to rank *all* the candidates on their ballot?

2. Let's keep using the example elections in Exercise #1 in Section 2.5 on page #66. This time, do the following for *all three elections:*

   (a) Determine the winner, and then the group ranking, using the Vote for Two method, given in Definition 4.3. (When you've narrowed it down to two candidates, use Majority Rules instead of Vote for Two.)

   (b) Then, find the winner using the Vote for Three method.

   (c) Compare the results to what you found for the Borda Count and the Nauru Score Method. Describe anything interesting you notice.

   This time, why do you think we asked you to do all this for Election 3, as well, even though some ballots were *not full?* Do you think any voters might have submitted different rankings if they knew that the Vote for Two (or Three) method would be used?

3. Back in Section 3.2, we wrote that some voting methods essentially ask for a Condorcet winner and then, when there isn't one, do *something else*. One such example is **Black's Method**, where that "something else" is *apply the Borda Count*. [67]

(a) Apply Black's Method to several of the examples in this chapter and in the exercises above this. That is, in any examples where there was *no Condorcet winner*, identify the Borda Count winner; that candidate would be the winner under Black's Method.

(b) Keep in mind that Black's Method only applies the Borda Count when there is *no Condorcet winner* to begin with, so it's worth considering whether Black's Method could ever disagree with the outcome of the Borda Count. That is, could an election have a Condorcet winner who is *different* from the Borda Count winner?

If you think this is possible, identify some examples from the examples and exercises in this book, or construct an example or two of your own. Alternatively, if you think this is impossible, provide a logical explanation of why that's the case.

4. This exercise is based on the results of a figure skating event at the 1994 Winter Olympics in Lillehammer, Norway. This event is described on page 22 of Donald G. Saari's book *Chaotic Elections*, where you will find the following table of ballots [52]:

	V1	V2	V3	V4	V5	V6	V7	V8	V9
B	1	1	1	1	1	2	2	3	3
K	2	2	2	2	2	1	1	1	1
L	3	3	3	3	3	3	3	2	2

The three competitors in question were Oksana Baiul (B), Nancy Kerrigan (K), and Chen Lu (L), and each voter (V1 through V9) is one of the judges. This time, the ballots are displayed with one column per voter, and the numbers 1, 2, and 3 to indicate 1st, 2nd, and 3rd place rankings.

(a) Convert this information into a table of ranked ballots like we have been using, where each column is a possible ballot, the entries in a column are the candidates (B, K, L) in some order, and the column headings are the number of voters who submitted that ballot.

(b) Then, apply the Plurality and Borda Count methods to those ballots to determine the winner in each case.

(c) Consider how you might have determined these same results from the original table above, before creating a table like the ones we've been using. Is one of them easier to work with than the other?

(d) In the actual event, Plurality was used. That is, only the judges' 1st place ranks determined the overall winner. Do you think this is a fair and reasonable method? Should the Borda Count have been used instead? In your response, consider what would happen if we *removed* Chen Lu from the competition and applied the Borda Count (which would amount to a Majority Rules contest with just two candidates). Isn't it strange that adding/removing that 3rd candidate L can possibly change whether $B > K$ or $K > B$ overall?

5. Look back to the exercise in Section 3.5 on page 110 with the ballots from the 2009 mayoral election in Burlington, VT.

(a) Determine the winner using the Borda Count.

(b) Do you think the voters may have acted any differently if they knew the Borda Count would be used instead of Instant Runoff?

(c) Can you *impose more information on the ballots* to produce a different winner? That is, what if we modified some of the non-full ballots to be full instead by assigning a 2nd and 3rd place rank? Can you do this in such a way that a different Borda Count winner emerges?

If this is possible, find at least one way to do so. If you think this is impossible, provide a logical explanation for why that's the case.

(d) Likewise, determine the winner using the Vote for Two method. Then, explain whether the voters may have acted differently if they knew that method would be used. And, determine whether or not imposing more information on the ballots could lead to a different Vote for Two winner.

(e) Finally, using the original ballots, determine whether different points lists could lead to each candidate possibly being the winner.

6. This exercise is based on the actual results of the 2021 election for the city council seat in Ward 2 of Minneapolis, Minnesota. There were three main candidates – Yusra Arab (A), Cam Gordon (G), and Robin Worlobah (W) – and the tables below show the 8,915 ranked ballots that we will use (after two minor candidates were already eliminated). [51]

# of voters	801	1177	822	908	756
1st	G	G	G	A	A
2nd	A	W	-	G	W
3rd	W	A	-	W	G

# of voters	1572	1299	1088	492
1st	*A*	*W*	*W*	*W*
2nd	-	*G*	*A*	-
3rd	-	*A*	*G*	-

(Note: These 9 columns were split into two tables merely to fit on the page. They collectively represent this one election.)

(a) Determine the winner using Plurality Voting. However, take note that "Worlobah was the actual plurality winner because she had the most first-place votes prior to the elimination of" the two minor candidates. [51] (By the way, how is that possible?)

(b) Determine the winner using Instant Runoff Voting. (This is what was actually used in the election.)

(c) Determine the winner using the Borda Count.

(d) Could you make the other candidates win by using a points list other than $[2, 1, 0]$? For each candidate, find a way to do so, or else explain why it's not possible.

(e) Then, determine the winner of each possible pairwise matchup. Is there a Condorcet winner or loser here?

(f) A scholarly article about this election also describes a *"downward monotonicity paradox"* in this example:

> *"If 80 of the voters who cast the ballot Arab, Gordon, Worlobah had instead moved Arab down on the ballot and voted Gordon, Arab, Worlobah, then Worlobah would be the first candidate eliminated and Arab would be the IRV winner."* [51]

Confirm those claims: rewrite the table of ballots with the proposed changes to those 80 voters, and then determine the winner using IRV.

(g) Based on all of these results, who do you think *"should"* win? If possible, come up with a convincing explanation for why Gordon could be the winner, then do the same for Worlobah and Arab.

7. Consult the Wikipedia page for the Borda Count, which has a section on *"Potential for tactical manipulation."* [68]

(a) That section contains an example of an election using the Borda Count. Take the information stated there and write the table of ranked ballots; then, apply the Borda Count to confirm the results.

(b) That section also describes how changes to the ballots could produce different winners, but the explanations are rather terse. So, create a detailed *"explainer"* that illustrates what those ballot changes are and how they lead to different outcomes.

Your goal is to be able to share whatever you create with friends, family, or colleagues to help them understand how the Borda Count may be susceptible to **strategic voting**.

8. Identify a scenario in your life where you have used something very similar to a ranked score method. (See Sections 4.2.1 and 4.2.2 for inspiration.) Or, identify a scenario in your life where you *could* do something like that, and describe why a ranked score method would be a reasonable choice.

9. *This exercise requires knowledge of **calculus**, specifically **integration**, and even integration with multiple variables.*

   Find and read the article "The Mean(est) Voting System" from the September 2016 issue of *Math Horizons* [69]. Describe the main result in your own words. How does it relate to the topics of this chapter?

   Then, create an *"explainer"* that illustrates the main result for the case with just $n = 3$ candidates. Your goal is to help a student in an integral calculus course understand (i) what the Borda Count and other ranked score methods are, and (ii) how integration can be used to find an *"average"* of some kind. (If you actually create an essay or presentation to share with your classmates, please contact us to tell us all about it!)

Chapter 5:
Cardinal Methods (not Ranked), including Approval Voting

A LL of the voting methods we have learned thus far are based on the voters ranking the candidates. We have seen several ways to incorporate those rankings into an overall result. Plurality Voting ignores all the rankings except for 1st place. Runoff methods look mostly at 1st place votes, but lower rankings get used as candidates are eliminated. Pairwise matchup methods use the rankings to compare candidates one-on-one, but they ignore the *distance* between their rankings. Ranked score methods seek to alleviate that shortcoming by awarding points to candidates based on how highly they are ranked on a voter's ballot.

You may be wondering if there are other voting methods that aren't based on ranked ballots. In other words, you may be wondering if there are other reasonable ways for voters to submit their preferences, and to have those individual preferences combined into an overall result. Indeed, there are, and the goal of this chapter is to introduce you to some voting methods that ask the voters to **score** the candidates instead of ranking them. We will call them **Cardinal Voting Methods** to distinguish them from all the **Ordinal Voting Methods** we have learned. ("Ordinal" means "based on order (ranking)," whereas "cardinal" means "based on size or number.") One of the more well-known examples of these methods is **Approval Voting**, and we will discuss some real world examples of that method.

This chapter will conclude this book about various voting methods and how they are used in the real world. After this, a followup book will guide you in investigating the mathematical and logical properties of all of the voting methods we have learned about here.

5.1 Definitions and Examples

5.1.1 Approval Voting

When you vote in a typical election in the US with the Plurality Voting method, the instructions on the ballot are: **"Vote for one."** There may be several candidates listed, and you are asked to select just one of them to support by filling in a bubble or something like that.

What if you want to support more than one candidate? Could you vote for several candidates, giving them one vote each? This wouldn't prevent other voters from choosing just their one favorite, and it would allow you to perhaps express your true preferences more accurately. This idea — **"Vote for one or more"** — is precisely what **Approval Voting** is all about.

Definition 5.1: Approval Voting.
Assume there is an election with any number of candidates. The main idea is to allow voters to select any number of candidates, and the candidate with the most votes wins.

*When a voter selects a candidate on their ballot, we say they are giving that candidate an **approval vote**, and this is worth **one point** towards that candidate's total. It does not matter how many candidates a voter selects: every approval vote is worth one point.*

157

The candidate with the most approval votes overall is declared the winner.

Figure 1: Comparing the instructions for a ballot using Plurality Voting (left) with a ballot using Approval Voting (right). Source: STL Approves [70]

By this point in the book, you may have expected that definition to say: "Assume voters submit *ranked ballots* (with no ties)." But we didn't say that! This is our first example of a voting method that is *not based on the voters ranking the candidates*. The next definition provides some terminology to clarify that distinction.

Definition 5.2: Cardinal versus Ordinal Voting Methods.
*An **ordinal voting method** is one where each voter submits a ranking of the candidates, and those rankings are used to decide the winner. ("**Ordinal**" means "pertaining to position in a sequence,", as in "a particular **order**.")*

*A **cardinal voting method** is one where each voter assigns each candidate a numerical value (a score) and those values are used to decide the winner. ("**Cardinal**" means "pertaining to numerical value," as in "**counting**.")*

You are already familiar with this distinction, but perhaps not the terminology. You already know there's a distinction between the number 1 (that's a **cardinal** number) and the idea of being 1st in a sequence (that's an **ordinal** number), even though they both involve the concept of "1."

Approval Voting is a cardinal voting method because each voter is assigning each candidate a score of either 1 or 0. If the voter selects a candidate, they're giving them an approval vote, which is worth 1 point. If the voter does not select a candidate, they're giving them 0 points.

Example 5.1. Let's look at a simple example of an election using Approval Voting. Along the way, we will introduce you to some aspects of this method that will inform our discussion of the method's pros and cons in Section 5.3.

Suppose the Mathematics Department has 10 faculty members and they are voting on who to hire for a new position. There are four candidates, and the department decides to use the Approval Voting method: each voter (faculty member) will select *as many of the candidates as they wish* on their ballot. The table below summarizes the ballots: each column is one voter (V1, V2, ...), each row is one candidate (A, B, C, D), and an X indicates a voter approving of that candidate (while a blank indicates *not* approving).

	V1	V2	V3	V4	V5	V6	V7	V8	V9	V10
A	X	X		X	X	X	X		X	X
B		X		X	X	X		X		
C		X	X		X	X	X		X	X
D		X	X	X						X

Overall, Candidate A is the winner. Here are the full results:

- Candidate A receives 8 approval votes from 10 voters, for an approval rating of 80%, because $\frac{8}{10} = 0.80$.

- Candidate B receives 5 approval votes from 10 voters, for an approval rating of 50%, because $\frac{5}{10} = 0.50$.

- Candidate C receives 7 approval votes from 10 voters, for an approval rating of 70%, because $\frac{7}{10} = 0.50$.

- Candidate D receives 4 approval votes from 10 voters, for an approval rating of 40%, because $\frac{4}{10} = 0.40$.

We mentioned the **approval rating** of each candidate, which is the number of approval votes they receive divided by the number of voters (which is the maximum number of approval votes one could possibly receive). Frequently, these values are communicated as a percentage. This is one of the advantages of the Approval Voting method: candidates can get a better understanding of how broad their support is.

Under Plurality Voting, where every voter can only support *one* candidate, the percentages of total votes would all have to add up to 100%. This time, though, notice that $80\%, 50\%, 70\%, 40\%$ collectively add up to way more than 100%. This is because many voters selected more than one candidate. So, the candidates (and the voters) can better see how much support they had from the electorate, instead of having to perfectly divvy up 100 percentage points.

Two voters (V1 and V8) happened to support only one candidate, just like Plurality Voting would require. This is perfectly acceptable! Nothing in the rules of Approval Voting say that a voter *must* choose more than one candidate, if they don't want to.

However, one voter (V2) happened to support *all* the candidates. This is not *illegal* under the Approval Voting rules, but it is perhaps a strange thing for that voter to do because *it does not affect the overall outcome* of the election. For comparison's sake, think about removing that voter (V2) from the process. Every single candidate's total score of approval votes would decrease by one: A

would still win, but with only 7 votes instead of 8 (with B getting 4, C getting 6, and D getting 3). In that sense, the results would be the same. But, we should be careful and say that *the approval ratings would change slightly*. For example, A would have an approval rating of $\frac{7}{9}$, or roughly 77.8%. That's a little bit less than the 80% rating they had when all 10 voters were included.

A similar effect occurs when a voter participates in the election but they leave their ballot blank, approving of *none* of the candidates. (This may be called **abstaining**, or an **abstention vote**.) Then, the overall outcome of the election would be the same, in terms of approval *votes*, but every candidate's approval *rating* would be a little bit *higher* when that voter is ignored.

Put yourself in the position of the leader of this department trying to make this decision. Having all of this information might be very helpful, much more so than only having each voter's *top choice*. It also may help you to break any ties that occur, because you can look at where each candidate's support came from. Did many voters select *only* that candidate, or did they get broader support? And, if a candidate drops out of the race (as is common in job searches), it's easy to just ignore their approval votes and move to the next best candidate. Moreover, there were 24 approval votes, in total, from just 10 voters, indicating that many voters approved of two or even three candidates. This shows that, overall, this was a good *pool of candidates*, which may provide useful, real-world contextual information for the person overseeing this election. These observations, and more, support the idea of at least considering the Approval Voting method in a real world context like this one.

5.1.2 Score Voting

Approval Voting is the simplest version of the general idea of allowing each voter to assign a numerical **score** to each candidate. We say "simplest" because it only allows the scores 1 and 0. The following definition generalizes this idea to broader ranges of scores.

Definition 5.3: Score Voting / Range Voting.
*Assume there is an election with any number of candidates. The main idea is to allow voters to assign each candidate a **score**, and the candidate with the highest total score wins.*

*The list of possible scores a voter can use is called the **range**, so this method is called both **Score Voting** and **Range Voting**. (We will use **Score Voting** in this book.) The ballots will make it clear what the range of allowed scores is.*

*Each candidate gets a **total score** by adding their individual scores from all the voters. The candidate with the highest total score is declared the winner. (This may also be reported as an average score, as long as any blank markings on the ballot are treated as the lowest possible score.)*

Governor Candidates	Score *each* candidate by filling a number (0 is worst; 9 is best)
1: Candidate A →	**(0)** (1)(2)(3)(4)(5)(6)(7)(8)(9)
2: Candidate B →	(0)(1)(2)(3)(4)(5)(6)(7)(8)**(9)**
3: Candidate C →	(0)(1)(2)(3)(4)(5)(6)**(7)**(8)(9)

Figure 2: An example of a ballot for the **Score Voting** method. The ballot indicates the **range** of scores — from 0 to 9 — and it specifies which endpoint means *least support* and which one means *most support*. Source: Wikipedia [71]

This should feel familiar because you're probably used to using score ratings like this. Have you ever rated something from one to five stars on an app store? Have you ever filled out a course evaluation for a professor? Then you've essentially used the idea of score voting!

Changing the range of scores doesn't really fundamentally change the voting method, by the way. For example, asking voters to score candidates from 0 to 9 is mathematically equivalent to asking them for scores from 1 to 10. Both of those ranges have ten possible scores. So, asking for scores from -4 to $+5$ would also be equivalent, too. However, we can understand the argument that those ranges may be *psychologically* different. (The possibility of assigning *negative points* may induce different behavior in the voters, as opposed to assigning *small but positive* point values.)

That comment only goes so far, though, because making the range of scores *narrower* or *wider* can have an impact. Let's say you use the 0 to 9 scale for three candidates, and you rate candidate A with 9 points, B with 7 points, and C with 3 points. If the method changes to Approval Voting – the narrowest range, with only 0 or 1 as your options – will you approve of both A and B? Or just A? It will depend on the context. But if the method had changed to, say, the 1 to 10 range, you'd have an easy decision: just add $+1$ to all your scores. So, when we said before that the range of scores doesn't really matter, we should have clarified that the *width* of the range certainly matters.

Example 5.2. Let's reconsider the situation in Example 5.1. Suppose the Math Department decided to use Score Voting instead with a range from 0 (worst) to 9 (best). The table below shows the ballots submitted by each of the ten voters (one per column). We will use these example scores to practice applying the Score Voting method and better understand its properties. For convenience, we have highlighted in blue any score that corresponds to an approval vote in the original election in Example 5.1. This is to help us better compare the results of these two "versions" of the same election. (Keep in mind: if this Score Voting election were happening in practice, the voters would probably *not* be thinking about approval votes while casting their ballots.)

	V1	V2	V3	V4	V5	V6	V7	V8	V9	V10
A	10	5	0	7	7	6	8	0	8	10
B	0	7	4	7	9	6	3	8	0	3
C	0	10	10	1	7	9	10	5	10	8
D	0	8	8	10	4	4	3	5	6	10

Before we analyze these results, we should point out (in case you don't notice) that these ballots are *consistent with the Approval Voting ballots* from that earlier example. For instance, look at voter V5: they scored Candidate *A* with a 7, and they approved of Candidate *A* in Example 5.1. So, it would only make sense that Candidates *B* and *C* get a score of 7 or higher, because they were also approved by this voter in that example.

Across the board, for each individual voter, if they approved of *someone*, then anyone else they approved must receive an equal or higher score than that *someone*. (In the real world, this may not be completely accurate, though! Remember that voters are real people, and not everyone behaves the same as everyone else, and not everyone even necessarily behaves in a logically consistent way with themselves, either.) So, before declaring the official election outcome, here are some observations:

- *Voters have different thresholds for approval.* Voter V2 approved of candidate *A* with a score of 5, and yet voter V9 did not even approve of candidate *D* with a score of 6. Each voter has their own individual preferences and decisions.

- *The scores can imply a ranking, but not necessarily the other way around.* Look at voter V2: on a ranked ballot, they would likely put $C > D > B > A$, based on their scores here $(10 > 8 > 7 > 5)$. That is, we can *infer* that ranking from their scores.

  However, that doesn't necessarily work the other way around. If we had held a ranked ballot election first (using the Borda Count method, let's say), and voter V2 submitted $C > D > B > A$, then would we be able to predict their scores in this election? Maybe not! Perhaps they would give A only 2 points instead of 5. Perhaps they would give C and B both 10 points, and they only ranked $C > B$ that way because they were forced to pick *one* to be in 1st place (because tied ranks are not allowed).

  In other words, the Score Voting method *allows for ties* (e.g. V10 gave 10 points to both A and D) and it allows the voters to express the *intensity of preferences* (e.g. V1 really likes *only* candidate A). This is the significant difference from ranked ballot methods.

- *Score Voting "contains" other voting methods.* To see what we mean, look at voter V1: they clearly want only candidate A to win and no one else. In a Plurality Voting election, this voter would surely vote for A; in some election with a ranked ballot, they would surely rank A 1st (and they might leave the rest of their ballot blank, submitting a *non-full ballot*).

  In other words, Score Voting allows this voter to "do everything they could before," and it even gives them more ability to *express the intensity of their preferences*. This is one of the main arguments in favor of Score Voting: it does not limit the voters' ability to express their preferences and, if anything, it expands that ability.

Alright, now we're ready for the results. The Score Voting method says to add the total scores for each candidate and compare them. Here, we will also report the *average score* for each candidate (dividing their total by 10, the number of voters) so that you can compare them, although this is not required. We'll skip the calculations and trust that you can check them.

- *C wins with 70 points*, an average score of 7.

- *A finishes 2nd with 61 points*, an average score of 6.1.

- *D finishes a close 3rd with 58 points*, an average of 5.8.

- *B finishes 4th with only 47 points*, an average of 4.7.

And here are some observations about those results:

- *These are **not** the same as the Approval Voting results!* Remember that A had the most approval votes with 8, and C was next with 7. However, C had a higher total score this time.

  More specifically, notice that C had many 10s, whereas A had several approval votes that corresponded to lower scores (5, 6, 7). In other words, the supporters of C were mostly strong supporters, whereas several supporters of A were only moderate supporters. Furthermore, C had only a single 0 score (from the voter who only liked A), whereas A had two 0s.

  (If you compare candidates B and D, you'll notice a similar phenomenon occurred between the Approval and Score Voting results.)

- *Perhaps the **median score** would be a better measure of support.* Take the ten scores given to Candidate C and list them in increasing order: 0, 1, 5, 7, 8, 9, 10, 10, 10, 10. The middle items of that list are the 8 and 9 (because there's an even number of items, there's no perfect middle). Their average is 8.5, and so that is the **median** score for candidate C. That's a pretty high score! It shows that half of the electorate gave C an 8.5 or higher, which shows *broad and strong support*.

  We point this out because some variants of Score Voting actually use the **median** (middle value) instead of the **mean** (total points / # of voters) for each candidate to determine the overall results.

- *Gathering this information can be helpful for whoever oversees the election.* Put yourself in the shoes of the Chair of the Math Department as they try to make a decision. If Plurality Voting had been used instead, 10 votes would have to be split among 4 candidates, leading to a potentially "decisive" but not-so-helpful result of, say, 4-3-2-1, or even 3-3-2-2. The Chair may get a sense of how her colleagues feel about the candidates from group discussions, but having those Score Voting ballots can be much more useful. They can be used to decide on the winning candidate (or two, if there are two openings), to make an informed decision to break a close tie, or figure out who to make an offer to if the leading candidate drops out.

5.1.3 Variations on Score Voting

There are some interesting variations on Score Voting. They're noteworthy not for changing the range of scores, but rather *what they do with the scores* to decide on the overall winner. We will specifically define one such method below and then apply it to the example above.

Definition 5.4: STAR Voting ("Score Then Automatic Runoff"). *Assume there is an election with any number of candidates. The main idea is to allow voters to assign each candidate a **score**, and a runoff is held between the candidates with the highest total scores.*

*The election has an established **range** of scores the voters can use, just like **Score Voting**. Each candidate gets a **total score** by adding their individual scores from all the voters. The two candidates with the highest total scores advance to a runoff round.*

*Let's say the two candidates that advance to the runoff are X and Y. In that runoff round, each voter's ballot is identified as favoring X (if X's score is higher), or favoring Y (if Y's score is higher), or neutral (if the voter gave X and Y the same score). The winner is the one who is favored by more voters. (That is, there is essentially a **majority rules** runoff round, using the scores on the original ballots.)*

Example 5.3. Let's apply this **STAR Voting** method to the ballots from Example 5.2. We already calculated the total scores: C has 70 points, A has 61 points, D has 58 points, and B has 47 points. This means C and A will advance to the runoff round because they are the top two candidates.

In that runoff round, we need to determine whether each voter *favored A or C*. Looking at the ballots, we see that C was favored most often:

- 3 voters preferred A: voters V1, V4, V10.

- 6 voters preferred C: voters V2, V3, V6, V7, V8, V9.

- 1 voter was neutral: voter V5.

Therefore, C would be declared the winner of this election using **STAR Voting**.

This may not be too surprising, because we already saw in Example 5.2 that C won by getting mostly high scores (including several 10s), even though A had more approval votes (in Example 5.1). However, we should mention that it is certainly possible for the result of STAR Voting to be different from the result of Score Voting: that is, it's possible to have an election where A and C get the two highest scores, with C the higher of the two, and yet A wins in the runoff round comparing just A and C. We recommend that you try to construct such an example! It can be very instructive to create something like that.

Before moving on, we'll mention some other variations on Score Voting:

- In a real world context, you may want to use the STAR Voting idea but actually hold a separate runoff election. That is, use the Score Voting ballots to identify the top two candidates, and then actually ask the voters to choose one of the two, regardless of how they voted in the first round.

- As we mentioned during Example 5.2, you may want to use the **median** score instead of the **average (mean)** score.

- Related to that, you may wish to allow voters to submit *"No Score"* for a candidate, something *different than 0*. In other words, perhaps a voter can decide to not score a candidate at all, just like we might allow a voter to submit a *non-full ranked ballot* in a real world election. This is what we meant by the final comment in Definition 5.3: those *blanks/non-votes* could be counted as 0s towards a total score, but they could also be *ignored entirely* when calculating the average scores. In your real world context, it should be made clear to the voters what a *blank* will mean.

- For elections with *multiple winners* (like seats on a City Council), you can simply declare the top two total scores to both be winners, for instance. Or, you could use the STAR Voting method to find the 1st winner, then set that candidate aside and run the STAR Voting method again to find the second winner. (Notice that this concept is similar to the idea of a **group ranking** that we introduced in Definition 2.4.)

5.2 Real World Usage

5.2.1 Situations Similar to Score Voting

As we've mentioned before, it's likely that the general concept of **Score Voting** is the *most familiar* to the general public, although it is unlikely that many people have thought of it in terms of ballots and elections before. The following list mentions several real world situations that employ the concept of cardinal voting methods by allowing "voters" to submit scores for the "candidates."

1. **The Likert scale for surveys.** Student evaluations of teachers and customer satisfaction surveys at a restaurant or business are essentially "score voting elections." The voters are the students/customers, but there aren't really "candidates" or a "winner." Rather, this is done to help understand the voters' opinions and gather some *quantitative data* to analyze.

2. **Star ratings for films, books, apps, etc.** This is very similar to the item above, but these situations are probably even more common in the real world, so they're worth mentioning separately. Suppose you're shopping for something online

170

and you see a "score" based on customer ratings. This is based on an average of all the "votes" from customers, using the specified range, typically 1 star (worst) to 5 stars (best). You can usually see a **histogram** of the scores, too, which essentially shows you how many "voters" submitted each particular score. (See Figure 3 for an example.)

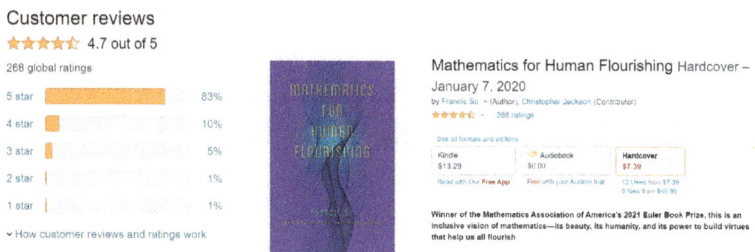

Figure 3: An example histogram of customer ratings for a product, which are essentially the scores submitted by "voters."

On the right is the main display for this product. Notice that the star rating is shown visually, without a specific numerical score. On the left is the histogram you would find after scrolling down quite a bit. There, you can even click on *"How customer reviews and ratings work"* to learn about how they incorporate customer ratings into the overall score.

Source: Amazon listing for *Mathematics for Human Flourishing*, Francis Su [72]

3. **GPA and Valedictorian.** Consider a typical method of choosing a valedictorian from a graduating class based on the highest grade point average (GPA). In that situation, each class a student takes allows the teacher to "vote" and assign the student a score (a grade of A counting as a 4.0, a B counting as a 3.0, etc.). Then, a student's GPA is like their *average score* in the "election," and the student with the highest av-

erage is the "winner." (Whether this is a good, fair method is a worthwhile discussion to have, but we'll leave that for another time.)

5.2.2   Real World Examples of Elections

There are a few places in the real world where Approval Voting or Score Voting, as we defined them in Section 5.1.1, are used. These seem to be mostly professional organizations and local political parties, but some recent ballot initiatives have brought cardinal voting methods to local politics, as well.

For example, Wikipedia mentions that the Green Party of Utah uses Score Voting, with a 0 to 9 range, to elect their officers. [73] The site also lists a few parties that use Approval Voting, like some local chapters of the Green and Libertarian Parties, as well as the German Pirate Party. [74]

The Democratic Party of Oregon used a modified version of STAR Voting to select delegates to send to the Democratic National Convention in 2020. (It is "modified" in the sense that there are multiple winners, and the party used the results to "elect a balanced number of male and female candidates, with non-binary candidates fairly included." [75]) Meanwhile, the Independent Party of Oregon's election for their Secretary of State in 2020 featured a wide variety of candidates, so they opted to use STAR Voting. A blog post analyzing the election results described it as "a serious stress test for STAR Voting" before concluding that it "successfully elected the candidate who best represented the electorate." [76]

It is worth pointing out that several national professional societies have all chosen to use Approval Voting as their method for electing officers. The Mathematical Association of America (MAA), the American Mathematical Society, the Institute of Management Sciences, and the Society for Social Choice & Welfare all use Approval Voting. [77, 78] Isn't that interesting, that all of these professional organizations devoted to mathematics, management, and social welfare have all chosen to use this method? This is certainly not reason *enough* to use Approval Voting in other situations, but it should at least give you something to think about. (Indeed, one of the first times this author ever heard of Approval Voting was using that method to vote for the President of the MAA [79], of which the author is a member.)

If you are interested in a through analysis of a specific election that used Approval Voting, we recommend Donald Saari's book *Chaotic Elections: A Mathematician Looks at Voting*. A section in Chapter 2 (pages 53-60) includes a thorough analysis of the 1999 election for the president of the Society for Social Choice & Welfare and, as part of that, a critique of the Approval Voting method. [52] (See the upcoming Section 5.3 for more comments.)

Finally, there have been some recent campaigns around the U.S. to adopt Approval or Score Voting at the local or state level. For example, voters in Fargo, North Dakota voted in November 2018 to use the Approval Voting method in the future. Then, in June 2020, voters used that method in races for City Commission, mayor, and municipal judge. Some local news reports have documented that process, how the voters feel about it, and whether the method is encouraging more candidates to run for office than usual. [80, 81, 82, 83, 84]

Other cities are also experimenting with Approval Voting and determining whether it is right for their contexts. For instance, voters in St. Louis overwhelmingly voted in favor of "Proposition D" in 2020 (68.15% to 31.85%), which mandates open, non-partisan primaries for offices like mayor and comptroller, followed by a runoff general election between just the top two candidates. [85] However, recent debates by the local Board of Aldermen indicate some support for further modifying that method. [86] Meanwhile, there is some activity in Denver to consider Approval Voting or Ranked Choice Voting in future local elections. [87, 88]

Overall, it's fair to say that Approval Voting and Score Voting are fairly new topics in public debates about election methods. It's important for these ideas to be studied more thoroughly, in terms of their theoretical, mathematical properties, as well as how they are implemented in the real world and how voters actually behave under these systems. A 2010 article in *The New Yorker* captures these ideas well in this quote:

> "Mathematics can suggest what approaches are worth trying, but it can't reveal what will suit a particular place, and best deliver what we want from a democratic voting system: to create a government that feels legitimate to people – to reconcile people to being governed, and give them reason to feel that, win or lose (especially lose), the game is fair. The novelty of range and approval voting in modern politics is so great that we can't know how they'll work out without running experiments." [89]

Indeed, one of the main goals of this book is helping you understand that there are *other ways to vote* besides Plurality Voting. We are not advocating for any particular method as the "best" in all situations. Rather, it behooves us all, as participants in a democratic society, to understand these systems and choose the one that fits best for our particular *context*, whatever that may be.

5.3 Discussion of Pros and Cons

5.3.1 Pros

- **Allows voters to express intensity of preferences.** This is the main comparison point between *ordinal (ranked)* methods and *cardinal (score)* methods. Ordinal methods do not allow voters to have candidates tied on their rankings, and they do not allow voters to express how far apart those rankings are. Score Voting at least allows voters to quantify the intensity of their preferences with numerical scores.

- **Satisfies the Montonicity property: boosting a score cannot hurt a candidate.** This *Monotonicity property* will be the subject of a chapter in this book's sequel, but the main idea is this: if a voter increases their score for a certain candidate, then that should not hurt that candidate's chances of winning. Believe it or not, we will see examples where voters moving a candidate *up* the ranking on their ballots actually causes that candidate to move *down* in the overall results. However, this kind of situation cannot occur with cardinal voting methods.

- **Familiarity and ease of use.** It's hard to discount the fact that assigning a score from within an allowed range is a process that almost everyone has done before (like customer ratings). If one of the main things keeping Plurality Voting in use is its simplicity, though, this point is notable, too.

- **Can convert to ranked ballots, but not the other way around.** We discussed this during Example 5.2, how the Score Voting ballots can be used to create corresponding

175

ranked ballots (although the issue of *ties* needs to be addressed). So, in practice, you may want to conduct a Score Voting election, then take the ballots submitted and convert them to ranked ballots, and then apply some other methods (like, say, the Borda Count) to them. This won't make sense in a political context, where clear rules need to be established about how the winner is chosen. However, in a more informal setting (like the leader of a club), you may want to do something like this to help you make a decision on behalf of a group.

- **Can make helpful information publicly available: approval ratings and average scores.** We also mentioned this during Examples 5.1 and 5.2. Unlike Plurality Voting, where the percentage of votes earned must total to 100%, Approval Voting allows for the voters and candidates to all learn about *how broad the support for each candidate is* by assessing their approval rating (or average scores, under Score Voting). In practice, this can help voters and candidates better understand the relevant political landscape and tailor their campaigns and policies.

- **Technical: Evades the conclusion of Arrow's Theorem.** This won't make sense right now, sorry! But, a major goal of this book's sequel will be to help you understand the most important and influential result in the mathematical study of voting methods: **Arrow's Theorem.** [90] In a nutshell, it says that all ordinal voting methods permit some kind of paradoxical results to occur. However, that theorem only applies to *ordinal (ranked) voting methods*, and not cardinal methods. So, one argument in favor of the methods in this chapter is simply that they seem to get around this significant theoretical road block.

5.3.2 Cons

- **Approval Voting can devolve to Plurality Voting, especially in a "close race."** This comment goes beyond the mathematics of the voting method itself and gets into the *behavior of the voters* and their strategies. So, we will have to keep the discussion short in the context of this book, but the main idea is this: in an election with two main front-runners, it's likely that Approval Voting will end up mimicking Plurality Voting, with many voters approving of just one of the two main candidates.

  Now, this may not seem like much of a criticism. But if the whole point of using Approval Voting is to avoid the disadvantages of Plurality Voting, then it is worth considering how often the method will play out "just like" Plurality Voting in the real world.

- **Reasonable behavior by individual voters can cause an undesirable outcome overall.** This is related to the previous point about the strategic behavior of the voters. And, in fact, this idea was of serious practical concern to the organizers when Approval Voting was used for the first time to elect officers for the Mathematical Association of America (MAA). Indeed, the president at the time issued an announcement with the election ballots:

  > "Suppose there are three candidates of whom two are outstanding. Suppose the third is a person you believe is not yet ready for office but whom

you decide to vote for as a means of encourage-
ment (in addition to voting for your favorite). If
enough voters reason that way, you will elect that
person now." [91]

Mind you, this is the organizer of an election cautioning the
voters about strategic voting *on the ballot itself!*

- **Relatively new, need more study and public discussion.**
  In general, cardinal voting methods are newer and less well-
  studied than ordinal voting methods. This is absolutely not a
  downside to the methods themselves, but we felt it important
  to mention this *somewhere*.

5.4   Practice Problems with Solutions

Practice Problem 5.1. This problem reconsiders the ongoing In-
dian/Pizza/Thai example using the methods from this chapter. Let's
say the nine friends decide to use the Approval Voting method, and
the submitted ballots are those shown below. (For convenience, we
are now using *one column per voter* so you can see everyone's
individual vote.)

Voters:	V1	V2	V3	V4	V5	V6	V7	V8	V9
Indian (I):	X	X	X	X					
Pizza (P):					X	X	X	X	
Thai (T):	X			X			X	X	X

Or, suppose instead that the friends used the range 0 (worst) to 10
(best) to score the dinner options instead, and the results are shown
below. (For convenience, any approval vote above is marked in blue
below.)

Voters:	V1	V2	V3	V4	V5	V6	V7	V8	V9
Indian (I):	10	10	9	8	0	8	2	5	2
Pizza (P):	5	3	0	2	10	10	9	7	4
Thai (T):	7	4	4	8	2	8	8	10	10

1. Explain how the first set of ballots (approval votes) can be seen as *consistent* with the ranked ballots from Example 2.3.

2. Likewise, explain whether the second set of ballots (range 0 to 9) can be seen as *consistent* with both the approval votes and the ranked ballots.

3. What are the results of the Approval Voting election?

4. What are the results using the standard Score Voting method?

5. What are the results using the STAR Voting method?

6. Suppose we use a variation of the STAR Voting method, but the runoff round compares the **median scores** of the candidates. What happens?

▼ **Solution:**

1. The approval votes at least match the 1st place votes from the ranked ballots. That is, it's reasonable to assume that a voter will approve of the candidate they ranked 1st.

   There are a few voters who also approved of their 2nd choice. For instance, voter V1 is one who ranked $I > T > P$, and they happened to also approve of their 2nd choice, Thai. Similarly, voter V8 is one who ranked $T > P > I$, and they happened to also approve of their 2nd choice, Pizza.

179

2. In each column, any time a voter approved of a candidate, any candidate whose score is *at least that big* should also be approved. For instance, voter V8 gave Pizza a score of 7, and they gave that option an approval vote. So, because they also approved Thai, the corresponding score should be ≥ 7. (And, indeed, it is a 10.)

However, there is one interesting "discrepancy" here. Look at voter V6: they gave a score of 8 to *both* Indian and Thai, meaning they are tied (or should we say *Thai-ed*?). This is *not allowed* using ranked ballots, so perhaps that voter truly was conflicted between those two options, and they just flipped a coin in their head to choose the ranking $T > P > I$ because they were forced to pick 2nd and 3rd choices. Voter V4 also had a tie in their scores. Other than those two instances, all other scores are consistent with the *strict* rankings from the ranked ballots.

3. Indian received 4 approval votes ($\frac{4}{9} = 0.444\ldots \approx 44.4\%$ rating). Pizza received 4 votes, as well. Thai received 5 approval votes ($\approx 55.5\%$ rating) and is declared the winner. How interesting that it was the *loser* under Plurality Voting, though!

4. Indian received a total of 54 points, Pizza received 50, and Thai received 60. So, this time, as well, Thai is declared the winner.

5. Indian and Thai had the top two scores, so they advance to the runoff. Only three voters gave Indian a higher score (V1, V2, V3), whereas four voters gave Thai a higher score (V5, V7, V8, V9). (The remaining two voters gave them tied scores.) So, Thai wins using STAR Voting.

6. There's another way to compare Indian and Thai one-on-one, and that's what this question is about. Although Indian received a lower total score (54 compared to 60), Indian has a *higher median score*.

  - Indian's scores in order: 0, 2, 2, 5, 8, 8, 9, 10, 10. The middle is the first 8: there are four scores below it and four above it. So, Indian's median score is 8.

  - Thai's scores in order: 2, 4, 4, 7, 7, 8, 10, 10. The middle is the second 7: there are four scores above and four scores below. So, Thai's median score is 7.

According to this new method, Indian would win the runoff.

How could this happen? Notice that more than half of Indian's scores are pretty high (8 and up) while the others are pretty low (0, 2, 2, 5). This makes the median pretty high, even though the total is pulled down by the 0 and the 2s. Meanwhile, Thai had no terrible scores (no 0s, a single 2), but they didn't have as many high scores as Indian did. This allowed Thai's total score to be higher, and yet their median is lower.

▲

5.5 Exercises

1. Consider the ballots (shown below) from Election 1 in Exercise #1 of Section 2.5. Our goal now is to explore how the results might change if Approval Voting is used instead.

# of voters	8	6	5	4	2
1st	B	A	C	D	C
2nd	C	C	A	A	D
3rd	D	B	D	B	A
4th	A	D	B	C	B

(a) Suppose that every voter chose to approve of just their 1st choice. Explain how this is identical to a voting method we already know.

(b) Suppose that every voter approves of their 1st choice, and half of the voters also approve of their 2nd choice. (For example, of the 8 voters in the first column, suppose 4 of them approve of just B, while 4 of them approve of B and C. Also, because the third column corresponds to an *odd* number (5) of voters, let's consider two possibilities: (i) three voters approve of just C while two approve of C and A, or (ii) three voters approve of C and A while two approve of just C.) What are the results this time, for both (i) and (ii)?

(c) By imposing approval votes on these ballots in other ways, could we make any of the four candidates win? For each one, demonstrate a way to make that candidate win, or else explain why that would be impossible while being consistent with the rankings on these ballots.

2. Similarly to the previous exercise, consider the ballots of Election 3 in Exercise #1 in Section 2.5 (reproduced below) and, for each of the four candidates, determine whether they could possibly win using Approval Voting, while being consistent with the rankings given.

# of voters	6	5	4	3	2	2	2	1	1	1
1st	C	A	B	C	A	D	B	A	D	B
2nd	B	D	A	A	D	B	A	B	-	-
3rd	-	C	D	D	B	-	-	-	-	-
4th	-	-	C	-	C	-	-	-	-	-

3. Let's revisit the example on page 150 that came from the 1994 Winter Olympics [52], whose ballots are reproduced below:

	V1	V2	V3	V4	V5	V6	V7	V8	V9
B	1	1	1	1	1	2	2	3	3
K	2	2	2	2	2	1	1	1	1
L	3	3	3	3	3	3	3	2	2

Recall that the candidates are Oksana Baiul (B), Nancy Kerrigan (K), and Chen Lu (L), that each voter (V1 through V9) is one of the judges, and that the table displays the judges' 1st, 2nd, and 3rd place rankings.

(a) Suppose that a cardinal voting method had been used instead, where the judges are allowed to assign each figure skater a score from 1 to 10, and all of those scores are used to determine a winner.

Assign scores to each judge's ballot in a manner that is consistent with their rankings. For example, judge V1 should assign B a higher score than K, and K a higher score than L. For simplicity, let's say that no judge assigned two skaters the same score.

Using those assigned scores you just made up, determine the winner using Score Voting and STAR Voting.

(b) If your example didn't already make this happen, then modify the scores so that two different winners arise from Score Voting and STAR Voting. Or, if you think this is impossible to accomplish with these ballots, provide a logical explanation as to why this is the case.

(c) Is it possible to make Chen Lu the winner using any of the cardinal voting methods in this chapter? Either provide an example of scores that make this happen, or else explain why this is impossible.

4. In Practice Problem 5.1 on page 178, we imposed some approval votes and scores on the ranked ballots from the Indian/Pizza/Thai election. These votes and scores led to Thai being the winner under the Approval, Score, and STAR Voting methods. Moreover, Thai also won when we used the *median* score in the STAR runoff round, instead of a pairwise comparison.

In this exercise, we want you to modify the scores in that example so that the Score and STAR methods yield different winners. Make sure you keep the scores consistent with the original rankings and the approval votes we already imposed, too! Either demonstrate how this can happen, or else provide a logical explanation for why this is impossible.

5. This exercise explores whether the results of an election using STAR Voting could be determined in separate districts, or whether the ballots need to be centrally combined before any calculations.

The tables below show the ballots submitted by 18 total voters, 9 in each of two districts. Each column corresponds to one possible ballot, the entries in the column show the scores for each of the five candidates (A to E), and the column headings state how many voters submitted those scores. (For instance, in the West District, three voters gave A 5 points, B 5 points, C 3 points, and D and E 0 points each, whereas one voter gave A 5 points, B and C 3 points each, and D and E 0 points each.)

West District:

# of voters	5	3	1
A	3	5	5
B	3	5	3
C	4	3	3
D	0	0	0
E	5	0	0

East District:

# of voters	5	3	1
A	3	5	5
B	0	0	0
C	4	3	3
D	3	5	3
E	5	0	0

(These ballots are taken from an example on the website RangeVoting.org [92], and the questions below are inspired by the discussion of that example.)

(a) Apply the STAR Voting method to the West District.

(b) Apply the STAR Voting method to the East District.

(c) Based on those results, make a prediction for what will occur when we combine the two districts into one larger election. Who do you think will win, using STAR Voting?

(d) Now, actually apply the STAR Voting method to all 18 ballots combined. What do you notice? Looking back at who advanced to the runoff round in each district, and who advanced to the runoff round in the combined election, identify how this phenomenon occurred.

(e) Finally, let's consider whether the standard Score Voting method is susceptible to this kind of paradoxical outcome. In this example, A won each individual district under Score Voting (considering each runoff round as a tie-breaker), and they were the overall Score Voting winner of the combined election.

Will this happen in *every* example? If candidate A wins under Score Voting in both districts (whether or not a tie-breaker runoff is needed), will that lead to A winning under Score Voting when the two districts are combined? If you think this is not true, construct a counterexample to show how A could possibly lose in the combined election. Or, if you think this is true, provide a logical explanation.

6. This exercise considers whether adding or removing a voter or candidate might change an outcome. The ballots are also taken from an example on `RangeVoting.org` [92], and the questions are inspired by the discussion there.

# of voters	4	3	2
A	3	3	3
B	2	5	1
C	4	0	4
D	5	0	0
E	0	4	5

(a) Determine the winner using Score and STAR Voting.

(b) Suppose the election organizers accidentally forgot to include one of the ballots where a voter assigned these scores: $A = 5$, $B = 0$, $C = 2$, $D = 0$, $E = 0$.

Knowing the results from (a) and looking at this voter's ballot, make a prediction for what will happen under Score and STAR Voting when this ballot is included in the election.

(c) Now, actually determine the winner using Score Voting and STAR Voting with all 10 voters (the original 9 and this forgotten ballot). What do you notice? Can you identify *how* this phenomenon occurred? What's the fundamental difference between standard Score Voting and STAR Voting that is allowing this to occur?

(d) Now, go back to the original 9 ballots. Suppose that, at the last minute, candidate B drops out of the race because they can no longer commit to the duties of the office. Make a prediction for how this will affect the results under Score and STAR Voting, then actually determine the winner using those methods. What do you notice?

(e) Summarize your findings from this exercise in a few sentences. What can we say about the possible effects of adding/removing a voter? Might that influence the results under Score Voting or STAR Voting? Likewise, what can we say about adding/removing a candidate?

7. Identify a political election in your area (a local, state, or federal office) that currently uses some kind of ordinal voting method (including Plurality). Write an op-ed that advocates for using some kind of cardinal voting method instead. As much as possible, support your opinions with concrete evidence and other ideas discussed in this book.

8. For this final exercise, we'd like you to actually *conduct an election* and analyze the results using all the voting methods we have learned about!

   (a) Identify a scenario in your life where you could reasonably ask voters to submit *scores* for the candidates. Maybe this will be an actual election for officers for a student club, or maybe this will be something just for fun, like family members voting for their favorite board game to play. Whatever it is, make sure you have *at least three candidates* and a reasonable number of voters. (The more voters there are, the more interesting the results might be, but you may need to use a spreadsheet to keep track of the results.) You may also decide what the range of scores will be, perhaps 0 to 5, or 0 to 9.

   (b) Apply the standard Score Voting method to determine a winner.

   (c) Apply the STAR Voting method to determine a winner.

   (d) Then, apply the Approval Voting method to determine a winner.

   Perhaps you want to ask voters to simultaneously score *and* approve of the candidates on their original ballot. (For example, the prompt could be: "Assign a score from 0 to 9 to each candidate. In addition, circle any candidate that you would *approve* of as a winner.")

   Or, perhaps you only ask voters for their scores and then impose the approval votes afterwards. (For example, if the range of scores was 0 to 9, you might assume that any score of 5 or higher amounts to an approval vote, because that's strictly above the median score of 4.5.)

188

(e) Next, use the voter's scores to establish *ranked ballots*. (We discussed this in Example 5.2.) We will leave it up to you how to deal with any *ties*; there is no "correct method" that should always be used.

Then, using those ranked ballots, apply some of the voting methods we learned in earlier chapters: Plurality, Instant Runoff, Top Two Runoff, Copeland's Method, and the Borda Count.

(f) Compare the results of all of these voting methods. Which one (or more) of them do you feel most accurately reflects *"the preferences of the electorate"* in your scenario?

(g) If it makes sense in your context, consider creating a presentation or explainer about all of these outcomes. The goal is to teach the voters who participated in your election about these voting methods and how they can be used to make reasonable and fair decisions on behalf of a group. And, as always, if you use this to actually make a decision in your life, please contact us to tell us all about it!

Conclusion

This concludes our introduction to various voting methods. We have now seen several different ways to use the information on ballots to determine the winner of an election, and we discussed how these voting methods are used in the real world. In various examples throughout the text, as well as in the practice problems and exercises, we have used our critical thinking and quantitative reasoning skills to explore how these voting methods work and better understand their mechanics. In addition to enhancing your general problem-solving skills, we'd like to think you've even picked up some practical advice about how to conduct elections in your own life: selecting officers for a student club, or settling disputes among friends and family, or just about any situation where a group needs to combine individual preferences to make a collective decision.

You may have already noticed the following two things about this book, but we're going to mention them now because they're important and closely related:

1. We have not advocated for any particular voting method as the "best" to use. Certainly, we haven't said that any one voting method is better than the rest in *all* situations. And, for any specific situation, we don't believe we've said that a certain voting method is "ideal" all the time. We've always said that *it depends on context.*

2. Based on the explorations we did in all the examples, practice problems, and exercises, it sure seems like every voting method allows for something "kinda weird" to happen. With some methods, we saw that boosting a candidate up some rankings caused them to lose the election instead. With others, we saw that taking away a candidate (or adding one in) at the last second could affect the outcome. And with others, we saw that counting the ballots and determining an outcome in separate "districts" might lead to different outcomes than if the ballots were combined into one larger election. On top of all that, we saw plenty of examples where several different candidates could reasonably be declared "the winner," depending on which voting method is used.

Item #1 is an important message to take away from this book: **there is no "perfect" voting method**. There is no way to change the rules of elections and just "solve" all the issues that inevitably arise when a group needs to make a collective decision in a reasonable, fair, and efficient manner. But that doesn't mean we should just accept whatever rules we currently have, as if *they* were inevitable. We have to be open to discussion, and so we must be aware of these potential options for making group decisions. We should know at least a little bit about how they work and what their pros and cons are. And we're not just talking about political elections here: these ideas could help improve how you select faculty committees, or host a baking competition, or choose the next common read for your book club, or anything else!

Still, as much as we *say* there is no "perfect" voting method, it's another thing altogether to *prove* that as a mathematical fact. That will be the goal of this book's sequel: to use abstract, mathematical reasoning to verify that intuition we described in Item #2. There is a real sense in which **"all voting methods can make weird stuff**

happen sometimes." Our goal will be to make that statement more precise and explore how true it is.

In the meantime, we hope you continue to learn about these ideas by using them in your own life or engaging in conversations about them. There's a lot to be gained by not just making decisions all the time, but also reflecting on *how* we make decisions.

Bibliography

[1] Fair Fight Action. Home | Fair Fight. `https://fairfight.com/`. Online; accessed 2022-08-13.

[2] The Young People's Project. About Us - The Young People's Project. `https://www.typp.org/about_us`. Online; accessed 2022-08-13.

[3] Wikipedia. First-past-the-post voting. `https://en.wikipedia.org/wiki/First-past-the-post_voting#Countries_using_FPTP/SMP`. Online; accessed 2022-05-30.

[4] Ballotpedia. Massachusetts' 3rd congressional district election (September 4, 2018 democratic primary). `https://ballotpedia.org/Massachusetts%27_3rd_Congressional_District_election_(September_4,_2018_Democratic_primary)`. Online; accessed 2022-05-31.

[5] Eric Sanders and Aaron Hamlin. What is Duverger's law? `https://electionscience.org/commentary-analysis/voting-theory-what-is-duvergers-law/`. Online; accessed 2022-05-30.

[6] Advance (MassLive.com). Massachusetts elections results, 2018-09-04. `https://elections.ap.org/ masslive/election_results/2018-09-04/ state/MA`. Online; accessed 2022-05-30.

[7] Secretary of the State (Connecticut) Denise W. Merrill. January 9, 2018 special election results 15th assembly district. `https://portal.ct.gov/-/media/SOTS/Ele ctionServices/ElectionResults/2018/15- Assembly-Results.pdf`. Online; accessed 2022-05-30.

[8] Ballotpedia. Recount laws in Connecticut. `https: //ballotpedia.org/Recount_laws_in_Conne cticut`. Online; accessed 2022-05-30.

[9] Dennis Hoey. Maine secretary of state's office publishes ranked-choice voting ballots. *Portland Press Herald*, 2018.

[10] Wikipedia. Australian House of Representatives. `https://en.wikipedia.org/wiki/Australian_ House_of_Representatives`. Online; accessed 2022-05-31.

[11] Francis Neely and Jason McDaniel. Overvoting and the equality of voice under instant-runoff voting in San Francisco. *California Journal of Politics and Policy*, 7(4), 2015.

[12] Francis Neely and Corey Cook. Whose votes count? undervotes, overvotes, and ranking in San Francisco's instant-runoff elections. *American Politics Research*, 36(4):530–554, 2008.

[13] FairVote. Ranked choice voting reduces ballot errors and inequities, new study shows. `https:`

//www.fairvote.org/ranked_choice_voting_ reduces_ballot_errors_and_inequities_ne w_study_shows. Online; accessed 2022-05-31.

[14] FairVote. Ranking is easy - a response to misleading claims about voter errors. https://www.fairvote .org/ranking_is_easy_a_response_to_misle ading_claims_about_voter_errors. Online; accessed 2022-05-31.

[15] City of Cambridge. Cambridge municipal elections. https://www.cambridgema.gov/Departments/ electioncommission/cambridgemunicipalele ctions. Online; accessed 2022-05-31.

[16] Wikipedia. Single transferable vote: History: United States. https://en.wikipedia.org/wiki/Single _transferable_vote#United_States. Online; accessed 2022-05-31.

[17] Wikipedia. 2017 French presidential election. https://en.wikipedia.org/wiki/2017_Fre nch_presidential_election#Results. Online; access 2022-05-30.

[18] Ballotpedia. Maine's 2nd congressional district election, 2018. https://ballotpedia.org/Maine %27s_2nd_Congressional_District_ele ction,_2018. Online; accessed 2022-05-30.

[19] Wikipedia. 2018 United States House of Representatives elections in maine. https://en.wikipedia.org/ wiki/2018_United_States_House_of_Represe

ntatives_elections_in_Maine#District_2.
Online; access 2022-05-30.

[20] Patrick Maloney. Electoral reform: City council votes 9-5 to scrap first-past-the-post voting and make London a Canadian trailblazer. *The London Free Press*, 2017.

[21] Wikipedia. President of India: Time of election: Election process. https://en.wikipedia.org/wiki/President_of_India#Election_process. Online; accessed 2022-05-31.

[22] FairVote. Details about ranked choice voting. https://www.fairvote.org/rcv#where_is_ranked_choice_voting_used. Online; accessed 2022-05-30.

[23] Ballotpedia. Presidential election in Maine, 2020. https://ballotpedia.org/Presidential_election_in_Maine,_2020. Online; accessed 2022-05-30.

[24] Ballotpedia. Alaska ballot measure 2, top-four ranked-choice voting and campaign finance laws initiative (2020). https://ballotpedia.org/Alaska_Ballot_Measure_2,_Top-Four_Ranked-Choice_Voting_and_Campaign_Finance_Laws_Initiative_(2020). Online; accessed 2022-05-30.

[25] FairVote. Rcv in campus elections. https://www.fairvote.org/rcv_in_campus_elections. Online; accessed 2022-05-30.

[26] FairVote. Rcv in private organizations and corporations. `https://www.fairvote.org/rcv_in_private _organizations_and_corporations`. Online; accessed 2022-05-30.

[27] FairVote. Oscar votes. `https://www.fairvote.org/ oscar_votes`. Online; accessed 2022-05-30.

[28] Sarah John, Haley Smith, and Elizabeth Zack. The alternative vote: Do changes in single-member voting systems affect descriptive representation of women and minorities? *Electoral Studies*, 54:90–102, 2018.

[29] Cynthia Richie Terrell, Courtney Lamendola, and Maura Reilly. Election reform and women's representation: Ranked choice voting in the US. *Politics and Governance*, 9(2):332–343, 2021.

[30] Represent Women. Research hub: Systems strategies: Ranked choice voting. `https://www.representwome n.org/research_voting_reforms#what_is_ right_with_rcv`. Online; accessed 2022-05-31.

[31] FairVote. Ranked choice voting elections benefit candidates and voters of color. `https://www.fairvote.org/re port_rcv_benefits_candidates_and_voters_ of_color`. Online; accessed 2022-05-31.

[32] FairVote. Data on ranked choice voting: Ranked choice voting and representation. `https://www.fairvote.org/ data_on_rcv#research_rcvrepresentation`. Online; accessed 2022-05-31.

[33] Robert Gehrke. Polling shows the public liked ranked choice voting, but Robert Gehrke explains why expanding it might be tough. *The Salt Lake Tribune*, 2021.

[34] Lee Drutman. Laboratories of democracy: San Francisco voters rank their candidates. it's made politics a little less nasty. *Vox*, 2019.

[35] Todd Donovan, Caroline Tolbert, and Kellen Gracey. Campaign civility under preferential and plurality voting. *Electoral Studies*, 42:157–163, 2016.

[36] Sean P. Thomas. Ranked-choice intrigue rises as Santa Fe mayor's race nears end. *Santa Fe New Mexican*, 2021.

[37] FairVote. Rcv civility project. https://www.fairvote.org/rcv_civility_project. Online; accessed 2022-05-31.

[38] Martha Kropf. Using campaign communications to analyze civility in ranked choice voting elections. *Politics and Governance*, 9(2):280–292, 2021.

[39] Eamon McGinn. Rating rankings: Effect of instant run-off voting on participation and civility. *Unpublished manuscript. Retrieved from http://eamonmcginn.com/papers/IRV_in_Minneapolis. pdf*, 2020.

[40] Sarah John and Andrew Douglas. Candidate civility and voter engagement in seven cities with ranked choice voting. *National Civic Review*, 106(1):25–29, 2017.

[41] RangeVoting.org. Example to demonstrate why IRV cannot be counted in precincts. https://rangevoting.org/IrvNonAdd.html. Online; accessed 2022-05-31.

[42] Ballotpedia. Maine's 2nd congressional district election, 2020. https://ballotpedia.org/Maine %27s_2nd_Congressional_District_ele ction,_2020. Online; accessed 2022-06-01.

[43] Wikipedia. History and use of instant-runoff voting. https://en.wikipedia.org/wiki/History_ and_use_of_instant-runoff_voting. Online; access 2022-06-01.

[44] Wikipedia. Ramon Llull. https://en.wikipe dia.org/wiki/Ramon_Llull#Other_Re cognition. Online; accessed 2022-05-31.

[45] M. Drton et al. The augsburg web edition of Llull's electoral writings. https://www.math.uni-augsburg.de /htdocs/emeriti/pukelsheim/llull/. Online; accessed 2022-05-31.

[46] Wikipedia. Schulze method. https://en.wikipe dia.org/wiki/Schulze_method. Online; accessed 2022-05-31.

[47] Pirate Party (US). Pirate national committee (pnc)/bylaws. https://wiki.uspirates.org/w/ index.php?title=Pirate_National_Committe e_(PNC)/Bylaws#Section_3:_Election. Online; accessed 2022-05-31.

[48] Gentoo Linux Wiki. Project:elections. https: //wiki.gentoo.org/wiki/Project:Ele ctions#Condorcet_method_of_voting. Online; accessed 2022-05-31.

[49] RangeVoting.org. An example of a range voting election in which cc (classic Condorcet) is disobeyed but wc (modified Condorcet property) is obeyed. `https://range voting.org/CondRangeExample.html`. Online; accessed 2022-05-31.

[50] Joseph T Ornstein and Robert Z Norman. Frequency of monotonicity failure under instant runoff voting: estimates based on a spatial model of elections. *Public Choice*, 161(1):1–9, 2014.

[51] David McCune and Lori McCune. The curious case of the 2021 Minneapolis ward 2 city council election. *arXiv preprint arXiv:2111.09846*, 2021.

[52] Donald G. Saari. *Chaotic elections!: A mathematician looks at voting.* American Mathematical Soc., 2001.

[53] Nancy Paczocha. Basic track & field meet knowledge. `http://p14cdn4static.sharpschool.com/Use rFiles/Servers/Server_819046/File/BASIC% 20TRACK%20MEET%20INFORMATION.pdf`. Online; accessed 2022-05-30.

[54] Russell Bruce Campbell. Team scoring. `https:// www.math.uni.edu/~campbell/mdm/score.html`. Online; accessed 2022-05-30.

[55] Wikipedia. Borda count: Current uses: Other uses. `https://en.wikipedia.org/wiki/Borda_ count#Other_uses`. Online; accessed 2022-05-30.

[56] Wikipedia. Ap poll. `https://en.wikipedia.org/ wiki/AP_Poll`. Online; accessed 2022-05-30.

[57] MarioKart Wiki. Driver's points. `https://mariokart.fandom.com/wiki/Driver%27s_Points`. Online; accessed 2022-05-30.

[58] Vivek Kaushik, Aubrey Laskowski, Matthew Romney, and Yukun Tan. 2015-2016 ap football poll using alternative voting methods. `https://demonstrations.wolfram.com/20152016APFootballPollUsingAlternativeVotingMethods/`. Online; accessed 2022-05-30.

[59] Wikipedia. Borda count: Current uses: Political uses. `https://en.wikipedia.org/wiki/Borda_count#Political_uses`. Online; accessed 2022-05-30.

[60] Thomas C Ratliff. Selecting committees. *Public Choice*, 126(3):343–355, 2006.

[61] Benjamin Reilly. Social choice in the south seas: Electoral innovation and the Borda count in the Pacific island countries. *International Political Science Review*, 23(4):355–372, 2002.

[62] Jon Fraenkel and Bernard Grofman. The Borda count and its real-world alternatives: Comparing scoring rules in Nauru and Slovenia. *Australian Journal of Political Science*, 49(2):186–205, 2014.

[63] Wikipedia. Jean-Charles de Borda. `https://en.wikipedia.org/wiki/Jean-Charles_de_Borda`. Online; accessed 2022-05-30.

[64] Wikipedia. Borda count: History. `https://en.wikipedia.org/wiki/Borda_count#History`. Online; accessed 2022-06-01.

[65] Janet Heine Barnett. The french connection: Borda, Condorcet and the mathematics of voting theory – voting in (and after) the revolution. *Convergence*, 2022.

[66] Photos en Images blog. Jean-Charles de Borda. `https://www.photosenimages.com/2016/10/jean-charles-de-borda.html`. Online; accessed 2022-05-30.

[67] Wikipedia. Black's method. `https://en.wikipedia.org/wiki/Black%27s_method`. Online; access 2022-06-02.

[68] Wikipedia. Borda count: Potential for tactical manipulation. `https://en.wikipedia.org/wiki/Borda_count#Potential_for_tactical_manipulation`. Online; access 2022-06-02.

[69] Sam Gutekunst, David Lingenbrink, and Michael E Orrison. The mean (est) voting system. *Math Horizons*, 24(1):10–13, 2016.

[70] STL Approves. Proposition D has been approved. `https://stlapproves.org/`. Online; accessed 2022-06-01.

[71] Wikipedia. Score voting. `https://en.wikipedia.org/wiki/Score_voting`. Online; accessed 2022-06-01.

[72] Amazon.com. Mathematics for human flourishing hardcover – january 7, 2020. `https://www.amazon.com/Mathematics-Human-Flourishing-Francis/dp/0300237138/`. Online; accessed 2022-06-01.

[73] Wikipedia. Score voting: Usage: Political use. https://en.wikipedia.org/wiki/Score _voting#Political_use. Online; accessed 2022-06-01.

[74] Wikipedia. Approval voting: Usage: Political organizations and jurisdictions. https://en.wikipedia.org/wiki/Approval_voting#Political_ organizations_and_jurisdictions. Online; accessed 2022-06-01.

[75] StarVoting.us. Democratic party adopts star for presedential delegate elections. https://www.starvoting.us/ star_dnc. Online; accessed 2022-06-01.

[76] STAR Voting. Independent party of Oregon star voting primary spotlight on the data: Almost 30% of voters showed no preference between the finalists in the secretary of state race. what does that mean for the results? *Medium.com*, 2020.

[77] Wikipedia. Approval voting: Usage: Other organizations and jurisdictions. https://en.wikipedia.org/wiki/ Approval_voting#Other_organizations. Online; accessed 2022-06-01.

[78] Steven J Brams and Peter C Fishburn. Approval voting in scientific and engineering societies. *Group Decision and Negotiation*, 1(1):41–55, 1992.

[79] Mathematical Association of America. Maa's internal election method meets with approval. https:// www.maa.org/news/math-news/maas-internal- election-method-meets-with-approval. Online; accessed 2022-06-01.

[80] Thomas Evanella. Fargo city commission election nation's first local contest to use approval voting. *INFORUM*, 2020.

[81] Robin Huebner. Approval voting to get its second major test with record slate of candidates in Fargo. *INFORUM*, 2022.

[82] Caitlyn Alley Peña and Chris Raleigh. You're darn tootin': Fargo just revolutionized American elections. *The Fulcrum*, 2020.

[83] Rob Port. Plain talk: Is approval voting in Fargo inspiring more candidates to run for office? *INFORUM*, 2022.

[84] Kelsey Piper. This city just approved a new election system never tried before in America. *Vox*, 2018.

[85] Ballotpedia. St. Louis, Missouri, proposition D, approval voting initiative (november 2020). `https://ballotpe dia.org/St._Louis,_Missouri,_Proposition_ D,_Approval_Voting_Initiative_(November_ 2020)`. Online; accessed 2022-06-01.

[86] Mark Schlinkmann. Effort underway to repeal 'approval voting' in St. Louis, replace it with new system. *St. Louis Post-Dispatch*, 2022.

[87] Meghan Lopez. Colorado and the city of Denver take a closer look at approval voting for the future of elections. *Denver 7*, 2021.

[88] David Sachs. How Denver's city elections might change. *Denverite*, 2022.

[89] Anthony Gottlieb. Win or lose: No voting system is flawless. but some are less democratic than others. *The New Yorker*, 2010.

[90] Kenneth J. Arrow. A difficulty in the concept of social welfare. *Journal of political economy*, 58(4):328–346, 1950.

[91] Alan D Taylor and Allison M Pacelli. *Mathematics and politics: strategy, voting, power, and proof.* Springer Science & Business Media, 2008.

[92] Warren D. Smith for RangeVoting.org. "STAR voting". `https://rangevoting.org/StarVoting.html`. Online; accessed 2022-06-02.